Elementary Zoology

MASTER BOOKS
CURRICULUM

Master Books Creative Team:
Authors: Gary E. Parker, Mary Parker, Bill and Merilee Clifton, Helen and Paul Haidle, Orit Kashtan
Editor: Craig Froman
Design: Terry White
Cover Design: Diana Bogardus
Copy Editors: Judy Lewis, Willow Meek
Curriculum Review: Kristen Pratt, Laura Welch, Diana Bogardus

First printing: March 2019
Fourth printing: November 2020

For information write:

Master Books®, P.O. Box 726, Green Forest, AR 72638
Master Books® is a division of the New Leaf Publishing Group, Inc.

ISBN: 978-1-68344-183-0
ISBN: 978-1-61458-706-4 (digital)

Unless otherwise marked, Scripture taken from the New King James Version.
Copyright © 1982 by Thomas Nelson, Inc. Used by permission. All rights reserved.

Printed in the United States of America

Please visit our website for other great titles:
www.masterbooks.com

Dr. Gary E. Parker is a popular homeschool author and speaker with multiple degrees, and a co-founder of Creation Adventures Museum. **Mary Parker** is a phenomenal amateur paleontologist who has participated in fossil digs around the world. **Bill and Merilee Clifton** are founders of Science Partners, offering creation-based science classes for homeschool students. **Helen and Paul Haidle** are an award-winning team who have created and published numerous books for children. **Orit Kashtan** lives in Israel with her husband, where both serve in leadership and ministry at the Grace & Truth Congregation. Both manage the HaChotam Christian Publishing house. *How Many Animals Were on the Ark?* includes a collection of authors from Answers in Genesis.

" Your reputation as a publisher is stellar. It is a blessing knowing anything I purchase from you is going to be worth every penny!

—Cheri ★ ★ ★ ★ ★

" Last year we found Master Books and it has made a HUGE difference.

—Melanie ★ ★ ★ ★ ★

" We love Master Books and the way it's set up for easy planning!

—Melissa ★ ★ ★ ★ ★

" You have done a great job. MASTER BOOKS ROCKS!

—Stephanie ★ ★ ★ ★ ★

" Physically high-quality, Biblically faithful, and well-written.

—Danika ★ ★ ★ ★ ★

" Best books ever. Their illustrations are captivating and content amazing!

—Kathy ★ ★ ★ ★ ★

Affordable
Flexible
Faith Building

Table of Contents

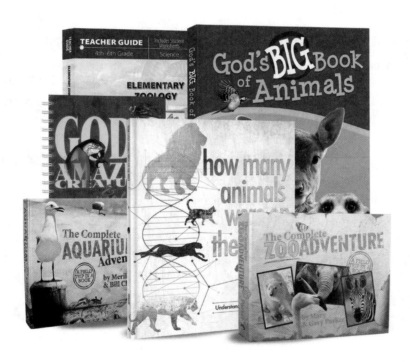

Using This Teacher Guide

Features: The suggested weekly schedule enclosed has easy-to-manage lessons that guide the reading, worksheets, and all assessments. The pages of this guide are perforated and three-hole punched so materials are easy to tear out, hand out, grade, and store. Teachers are encouraged to adjust the schedule and materials needed in order to best work within their unique educational program.

A Wild Adventure! Learn all about some of God's unique critter creations in this *Elementary Zoology* course! Students will study a variety of animals from mammals to birds and reptiles to fish, reading about their habitats, special characteristics, diets, cool facts, and more. With activities including dot-to-dots, coloring pages, and word searches, students will have tons of fun learning about animals from a biblical perspective. The course is designed for field trips to an aquarium and a zoo but also includes activities for an at-home animal adventure!

🕐	**Approximately 30 to 45 minutes per lesson, five days a week**
🔑	**Includes answer keys for worksheets, quizzes, and tests**
✏️	**Worksheets to help assess student learning**
📄	**Quizzes and tests are included to help reinforce learning and provide assessment opportunities; optional final exam included**
🔁	**Designed for grades 4 to 6 in a one-year science course**

Course Objectives: Students completing this course will

- ✔ Become familiar with the incredible range of life in God's wonderful world — from the oceans to the sky and everywhere in between
- ✔ Identify endangered species, their environments, and the importance of the natural world
- ✔ Learn about 45 unique creatures and important biblical principles in the fascinating devotional component of the course

- ✔ Study obscure animal facts, animal records, and amazing comparisons that highlight the uniqueness and variety found in God's creations
- ✔ Discover important information about animal kinds, Creation, and Noah's Ark

Course Description

Zoology is the study of life, often connected with biology and focused on animals, which is why we have those places called "zoos" all across the country. However, while the secular focus of zoology is on life origins and evolution, this course celebrates the wonder of God's creation and His unique design of all life on earth. This course has been developed to enhance learning about the diverse and amazing animals we see in our world. From the tiniest to the largest, you will be exploring what makes them unique as you celebrate their place in the world God created. Although the material can be modified for your own educational purposes, either on an elementary or more advanced level, it has been organized here to fit a one-year course.

Devotionals: Students who take this course will be reading one or two animal-focused devotions each Friday from the book *God's Amazing Creatures & Me!* Each gives the student information about an animal, a connected concept within the biblical text, a question that prompts personal reflection, and a memory verse. There is no testing or assessment for this element of the Elementary Zoology course.

Quizzes: Quizzes are optional and should be assigned at the teacher's discretion. The maturity of the student should determine whether the quizzes are open book.

Grading Options for This Course: It is always the prerogative of an educator to assess student grades however he or she might deem best. The following is only a suggested guideline based on the material presented through this course. To calculate the percentage of the worksheets, quizzes, and tests, the educator may use the following guide. Divide total number of questions correct (example: 43) by the total number of questions possible (example: 46) to calculate the percentage out of 100 possible. 43/46 = 93 percent correct.

The suggested grade values are noted as follows:

90 to 100 percent = A

80 to 89 percent = B

70 to 79 percent = C

60 to 69 percent = D

0 to 59 percent = F

Note: An aquarium visit and a zoo visit are discussed at the end of each semester. You might look through the last semester pages a few weeks prior to make sure everyone is ready for the trip. If you cannot get away for an aquarium or zoo visit, alternative ideas are provided to have a wonderful time at home no matter what.

Supply List

The following supplies are suggested for this course, including crayons, pencils, or markers for the pages that involve coloring:

How Many Animals Were on the Ark? – Worksheet 31: Measuring Activity

- ☐ yarn or string
- ☐ a measuring tape
- ☐ scissors

The Complete Aquarium Adventure – Worksheet 5: Water Cycle Bracelet

- ☐ pony beads of yellow, blue, green, white, clear, and brown
- ☐ elastic cord cut to 8 inches (or optional string or twine)

The Complete Aquarium Adventure – Worksheet 9: Guess Who? (Game)

- ☐ paper clips or clothespins

The Complete Aquarium Adventure – Worksheet 21: Top Marine Predator Mobile

- ☐ Styrofoam™, papier-mâché, balsa wood
- ☐ fabric, or other material light enough to suspend from the ceiling

The Complete Aquarium Adventure – Worksheet 24: Aquarium Diorama

- ☐ shoe box or similar cardboard box
- ☐ scissors
- ☐ glue
- ☐ tape
- ☐ plastic wrap
- ☐ construction paper or pictures
- ☐ additional items that might include sand, plants, and rocks

Bonus Activity: Green Sea Turtle Hand Puppet

- ☐ scissors
- ☐ tape

Bonus Activities: Sea Scramble 1 and 2

- ☐ scissors

First Semester Suggested Daily Schedule

Date	Day	Assignment	Due Date	✓	Grade
		First Semester-First Quarter			
Week 1	Day 1	Read *How Many Animals Were on the Ark?* • (HMAWOTA) Pages 4–5 • Complete Worksheet 1 • Page 17 • Teacher Guide (TG)			
	Day 2	Read Pages 6–7 • (HMAWOTA) Complete Worksheet 2 • Pages 19–20 • (TG)			
	Day 3	Creation Art • Complete Worksheet 3 • Page 21 • (TG)			
	Day 4	Read Pages 8–9 • (HMAWOTA) Complete Worksheet 4 • Page 22 • (TG)			
	Day 5	Read "Follow a Good Example" • Pages 6–7 • *God's Amazing Creatures & Me* • (GAC)			
Week 2	Day 6	Read Pages 10–11 • (HMAWOTA) Complete Worksheet 5 • Page 23 • (TG)			
	Day 7	Read Pages 12–13 • (HMAWOTA) Complete Worksheet 6 • Page 24 • (TG)			
	Day 8	Creation Art • Complete Worksheet 7 • Page 25 • (TG)			
	Day 9	Read Pages 14–15 • (HMAWOTA) Complete Worksheet 8 • Page 26 • (TG)			
	Day 10	Read "Deadly Beauty" • Pages 8–9 • (GAC)			
Week 3	Day 11	Read Pages 16–17 • (HMAWOTA) Complete Worksheet 9 • Page 27 • (TG)			
	Day 12	Read Pages 18–19 • (HMAWOTA) Complete Worksheet 10 • Page 28 • (TG)			
	Day 13	Creation Art • Complete Worksheet 11 • Page 29 • (TG)			
	Day 14	Read Pages 20–21 • (HMAWOTA) Complete Worksheet 12 • Page 30 • (TG)			
	Day 15	Read "No Two Stripes Are Alike" • Pages 10–11 • (GAC)			
Week 4	Day 16	Read Pages 22–23 • (HMAWOTA) Complete Worksheet 13 • Page 31 • (TG)			
	Day 17	Read Pages 24–25 • (HMAWOTA) Complete Worksheet 14 • Page 32 • (TG)			
	Day 18	Creation Art • Complete Worksheet 15 • Page 33 • (TG)			
	Day 19	Read Pages 26–27 • (HMAWOTA) Complete Worksheet 16 • Page 34 • (TG)			
	Day 20	Read "Don't Be A Quitter!" • Pages 12–13 • (GAC)			
Week 5	Day 21	Read Pages 28–29 • (HMAWOTA) Complete Worksheet 17 • Page 35 • (TG)			
	Day 22	Read Pages 30–31 • (HMAWOTA) Complete Worksheet 18 • Page 36 • (TG)			
	Day 23	Creation Art • Complete Worksheet 19 • Page 37 • (TG)			
	Day 24	Read Pages 32–33 • (HMAWOTA) Complete Worksheet 20 • Page 38 • (TG)			
	Day 25	Read "The Contented Cockroach" • Pages 14–15 • (GAC)			

Date	Day	Assignment	Due Date	✓	Grade
Week 6	Day 26	Read Pages 34–35 • (HMAWOTA) Complete Worksheet 21 • Page 39 • (TG)			
	Day 27	Read Pages 36–37 • (HMAWOTA) Complete Worksheet 22 • Page 40 • (TG)			
	Day 28	Creation Art • Complete Worksheet 23 • Page 41 • (TG)			
	Day 29	Read Pages 38–39 • (HMAWOTA) Complete Worksheet 24 • Page 42 • (TG)			
	Day 30	Read "The Perfect Home" and "Amazing Wings" • Pages 16–19 • (GAC)			
Week 7	Day 31	Read Pages 40–41 • (HMAWOTA) Complete Worksheet 25 • Page 43 • (TG)			
	Day 32	Read Pages 42–43 • (HMAWOTA) Complete Worksheet 26 • Page 44 • (TG)			
	Day 33	Creation Art • Complete Worksheet 27 • Pages 45–46 • (TG)			
	Day 34	Read Pages 44–45 • (HMAWOTA) Complete Worksheet 28 • Page 47 • (TG)			
	Day 35	Read "Watch That Tongue" • Pages 20–21 • (GAC)			
Week 8	Day 36	Read Pages 46–47 • (HMAWOTA) Complete Worksheet 29 • Page 48 • (TG)			
	Day 37	Read Pages 48–49 • (HMAWOTA) Complete Worksheet 30 • Page 49 • (TG)			
	Day 38	Creation Art • Complete Worksheet 31 • Pages 51–52 • (TG)			
	Day 39	Read Pages 50–51 • (HMAWOTA) Complete Worksheet 32 • Page 53 • (TG)			
	Day 40	Read "Listen Carefully" and "The King of Cats" • Pages 22–25 (GAC)			
Week 9	Day 41	Read Pages 52–53 • (HMAWOTA) Complete Worksheet 33 • Page 54 • (TG)			
	Day 42	Read Pages 54–55 • (HMAWOTA) Complete Worksheet 34 • Page 55 • (TG)			
	Day 43	Creation Art • Complete Worksheet 35 • Page 56 • (TG)			
	Day 44	Read Pages 56–58 • (HMAWOTA) Complete Worksheet 36 • Page 57 • (TG)			
	Day 45	Read "A Wise Builder" and "A Fearless Flier" • Pages 26–29 (GAC)			
	First Semester-Second Quarter				
Week 1	Day 46	Read Devotional 1 • Pages 12–13 • *The Complete Aquarium Adventure* • (TCAA) • Complete Worksheet 1 • Page 61 • (TG)			
	Day 47	Read Introduction to Birds (lefthand side) and Birds • Page 35 (TCAA) • Complete Worksheet 2 • Page 62 • (TG)			
	Day 48	Read Anhinga & Double-crested Cormorant • Pages 36–37 (TCAA) • Complete Worksheet 3 • Page 63 • (TG)			
	Day 49	Read Brown Pelican • Pages 38–39 • (TCAA) Complete Worksheet 4 • Page 64 • (TG)			
	Day 50	Read "Struggles That Strengthen" • Pages 30–31 • (GAC)			

Date	Day	Assignment	Due Date	✓	Grade
Week 2	Day 51	Read Devotional 2 • Pages 14–15 • (TCAA) Complete Worksheet 5 • Page 65 • (TG)			
	Day 52	Read Penguin • Pages 40–42 • (TCAA) Complete Worksheet 6 • Page 66 • (TG)			
	Day 53	Read Introduction to Fish (lefthand side) and Bony Fish • Page 43 • (TCAA) • Complete Worksheet 7 • Page 67 • (TG)			
	Day 54	Read Common Clownfish • Pages 44–46 • (TCAA) Complete Worksheet 8 • Page 68 • (TG)			
	Day 55	Read "Fight or Make Peace" and "Grace and Beauty" • Pages 32–35 • (GAC)			
Week 3	Day 56	Read Devotional 3 • Pages 16–17 • (TCAA) Complete Worksheet 9 • Page 69 • (TG)			
	Day 57	Read Great Barracuda • Pages 47–48 • (TCAA) Complete Worksheet 10 • Page 70 • (TG)			
	Day 58	Read Green Moray Eel • Pages 49–50 • (TCAA) Complete Worksheet 11 • Page 71 • (TG)			
	Day 59	Read Leafy Sea Dragon • Pages 51–52 • (TCAA) Complete Worksheet 12 • Page 72 • (TG)			
	Day 60	Read "A Deep Voice" • Pages 36–37 • (GAC)			
Week 4	Day 61	Read Devotional 4 • Pages 18–19 • (TCAA) Complete Worksheet 13 • Page 73 • (TG)			
	Day 62	Read Lionfish • Pages 53–54 • (TCAA) Complete Worksheet 14 • Page 74 • (TG)			
	Day 63	Read Long-spine Porcupinefish • Pages 55–56 • (TCAA) Complete Worksheet 15 • Page 75 • (TG)			
	Day 64	Read Seahorse • Pages 57–58 • (TCAA) Complete Worksheet 16 • Page 76 • (TG)			
	Day 65	Read "Dead…Or Alive?" and "The 'Blind' See" • Pages 38–41 (GAC)			
Week 5	Day 66	Read Devotional 5 • Pages 20–21 • (TCAA) Complete Worksheet 17 • Page 77 • (TG)			
	Day 67	Read Cartilaginous Fish • Page 59 • (TCAA) Complete Worksheet 18 • Page 78 • (TG)			
	Day 68	Read Bonnethead Shark • Pages 60–61 • (TCAA) Complete Worksheet 19 • Page 79 • (TG)			
	Day 69	Read Nurse Shark • Pages 62–63 • (TCAA) Complete Worksheet 20 • Page 80 • (TG)			
	Day 70	Read "The Potter Provides" and "The Need for Speed" • Pages 42–45 • (GAC)			
Week 6	Day 71	Read Devotional 6 • Pages 22–23 • (TCAA) Complete Worksheet 21 • Page 81 • (TG)			
	Day 72	Read Sand Tiger Shark • Pages 64–65 • (TCAA) Complete Worksheet 22 • Page 82 • (TG)			
	Day 73	Read Stingrays, Cownose Ray, and Southern Stingray • Pages 66–68 • (TCAA) • Complete Worksheet 23 • Page 83 • (TG)			
	Day 74	Read Invertebrates (lefthand side), Invertebrates and Coral Pages 69–71 • (TCAA) • Complete Worksheet 24 • Page 84 • (TG)			
	Day 75	Read "One Body" and "A Powerful Bird of Prey" • Pages 46–49 (GAC)			

Date	Day	Assignment	Due Date	✓	Grade
Week 7	Day 76	Read Devotional 7 • Pages 24–25 • (TCAA) • Complete Worksheet 25 • Page 85 • (TG)			
	Day 77	Read Giant Pacific Octopus and Horseshoe Crab • Pages 72–75 (TCAA) • Complete Worksheet 26 • Page 86 • (TG)			
	Day 78	Read Jellyfish, Moon Jelly, and Upside-down Jellyfish • Pages 76–79 • (TCAA) • Complete Worksheet 27 • Page 87 • (TG)			
	Day 79	Read Sea Anemone and Sea Star • Pages 80–84 • (TCAA) Complete Worksheet 28 • Page 88 • (TG)			
	Day 80	Read "Tiny, But Spiny" and "Listen For The Shepherd" • Pages 50–53 • (GAC)			
Week 8	Day 81	Read Mammals (lefthand side) and Cetaceans • Pages 85–89 (TCAA) • Complete Worksheet 29 • Page 89 • (TG)			
	Day 82	Read Bottlenose Dolphin and Killer Whale • Pages 90–93 (TCAA) • Complete Worksheet 30 • Page 90 • (TG)			
	Day 83	Read Beluga • Pages 94–95 • (TCAA) • Complete Worksheet 31 Page 91 • (TG)			
	Day 84	Read Pinnipeds, Harbor Seal, and California Sea Lion • Pages 96–102 • (TCAA) • Complete Worksheet 32 • Page 92 • (TG)			
	Day 85	Read "Slow Down!" • Pages 54–55 • (GAC)			
Week 9	Day 86	Read Reptiles (lefthand side) and Green Sea Turtle • Pages 103–105 • (TCAA) • Complete Worksheet 33 • Page 93 • (TG)			
	Day 87	Read Loggerhead Sea Turtle • Pages 106–107 • (TCAA) • Complete Worksheet 34 • Page 94 • (TG)			
	Day 88	Read American Alligator • Pages 108–111 • (TCAA) • Complete Worksheet 35 • Page 95 • (TG)			
	Day 89	**Take Quiz 1 - Aquarium Animal Crossword** (use names from page 114 TCAA) • Page 177 • (TG)			
	Day 90	Aquarium Field Trip or Aquarium in the House Day Page 187 • (TG)			
		Mid-Term Grade			

Second Semester Suggested Daily Schedule

Date	Day	Assignment	Due Date	✓	Grade
		Second Semester-Third Quarter			
Week 1	Day 91	Read Hummingbirds and Toucans • Pages 8–15 • *God's Big Book of Animals* • (GBBA) • Complete Worksheet 1 Page 99 • (TG)			
	Day 92	Read Pileated Woodpeckers and Crows • Pages 16–23 • (GBBA) Complete Worksheet 2 • Page 100 • (TG)			
	Day 93	Read Vultures and Owls • Pages 24–31 • (GBBA) Complete Worksheet 3 • Page 101 • (TG)			
	Day 94	Read Woodcocks and Seagulls • Pages 32–39 • (GBBA) Complete Worksheet 4 • Page 102 • (TG)			
	Day 95	Read "Where's The Map?" and "Friends Forever" • Pages 56–59 • (GAC)			
Week 2	Day 96	Read Geese and Swans • Pages 40–47 • (GBBA) Complete Worksheet 5 • Page 103 • (TG)			
	Day 97	Read Grebes and Pelicans • Pages 48–55 • (GBBA) Complete Worksheet 6 • Page 104 • (TG)			
	Day 98	Read Heron and Penguins • Pages 56–63 • (GBBA) Complete Worksheet 7 • Page 105 • (TG)			
	Day 99	Read Turkeys • Pages 64–67 • (GBBA) Complete Worksheet 8 • Page 106 • (TG)			
	Day 100	Read "'Bee' What You Can Be" and "Shout It Out!" • Pages 60–63 • (GAC)			
Week 3	Day 101	Read Monarch Butterflies and Moths • Pages 68–75 • (GBBA) Complete Worksheet 9 • Page 107 • (TG)			
	Day 102	Read Bees and Wasps/Hornets • Pages 76–83 • (GBBA) Complete Worksheet 10 • Page 108 • (TG)			
	Day 103	Read Mosquitoes and Flies • Pages 84–91 • (GBBA) Complete Worksheet 11 • Page 109 • (TG)			
	Day 104	Read Fleas and Termites • Pages 92–99 • (GBBA) Complete Worksheet 12 • Page 110 • (TG)			
	Day 105	Read "The Smallest Jewel" • Pages 64–65 • (GAC)			
Week 4	Day 106	Read Poison Dart Frogs and Turtles • Pages 100–107 • (GBBA) Complete Worksheet 13 • Page 111 • (TG)			
	Day 107	Read Alligators and Komodo Dragons • Pages 108–115 • (GBBA) Complete Worksheet 14 • Page 112 • (TG)			
	Day 108	Read Marine Iguanas and Chameleons • Pages 116–123 (GBBA) • Complete Worksheet 15 • Page 113 • (TG)			
	Day 109	Read Rattlesnakes • Pages 124–127 • (GBBA) Complete Worksheet 16 • Page 114 • (TG)			
	Day 110	Read "Time Changes Things"• Pages 66–67 • (GAC)			

Date	Day	Assignment	Due Date	✓	Grade
Week 5	Day 111	Read Deer and Camels • Pages 128–135 • (GBBA) Complete Worksheet 17 • Page 115 • (TG)			
	Day 112	Read Elephants and Gorillas • Pages 136–143 • (GBBA) Complete Worksheet 18 • Page 116 • (TG)			
	Day 113	Read Rabbits and Opossums • Pages 144–151 • (GBBA) Complete Worksheet 19 • Page 117 • (TG)			
	Day 114	Read Shrews and Mice • Pages 152–159 • (GBBA) Complete Worksheet 20 • Page 118 • (TG)			
	Day 115	Read "Warning! Watch Out!" • Pages 68–69 • (GAC)			
Week 6	Day 116	Read Squirrels and Groundhogs • Pages 160–167 • (GBBA) Complete Worksheet 21 • Page 119 • (TG)			
	Day 117	Read Beavers and Porcupines • Pages 168–175 • (GBBA) Complete Worksheet 22 • Page 120 • (TG)			
	Day 118	Read Skunks and Raccoons • Pages 176–183 • (GBBA) Complete Worksheet 23 • Page 121 • (TG)			
	Day 119	Read Badgers and Otters • Pages 184–191 • (GBBA) Complete Worksheet 24 • Page 122 • (TG)			
	Day 120	Read "Help Your Enemy" • Pages 70–71 • (GAC)			
Week 7	Day 121	Read Weasels and Meerkats • Pages 192–199 • (GBBA) Complete Worksheet 25 • Page 123 • (TG)			
	Day 122	Read Red Foxes and Wolves • Pages 200–207 • (GBBA) Complete Worksheet 26 • Page 124 • (TG)			
	Day 123	Read Lions and Tigers • Pages 208–215 • (GBBA) Complete Worksheet 27 • Page 125 • (TG)			
	Day 124	Read Grizzly Bears and Bats • Pages 216–223 • (GBBA) Complete Worksheet 28 • Page 126 • (TG)			
	Day 125	Read "Living On Little" • Pages 72–73 • (GAC)			
Week 8	Day 126	Read Dolphins • Pages 224–227 • (GBBA) Complete Worksheet 29 • Page 127 • (TG)			
	Day 127	Read Beluga Whales • Pages 228–231 • (GBBA) Complete Worksheet 30 • Page 128 • (TG)			
	Day 128	Read Atlantic Salmon • Pages 232–235 • (GBBA) Complete Worksheet 31 • Page 129 • (TG)			
	Day 129	Read Great White Sharks • Pages 236–239 • (GBBA) Complete Worksheet 32 • Page 130 • (TG)			
	Day 130	Read "Rejoice Always" • Pages 74–75 • (GAC)			
Week 9	Day 131	Read Octopus • Pages 240–243 • (GBBA) Complete Worksheet 33 • Page 131 • (TG)			
	Day 132	Read Jellyfish • Pages 244–247 • (GBBA) Complete Worksheet 34 • Page 132 • (TG)			
	Day 133	Animal Book Color Page Complete Worksheet 1 • Page 135 • (TG)			
	Day 134	Animal Book Color Page Complete Worksheet 2 • Page 136 • (TG)			
	Day 135	Read "Be A Blessing" • Pages 76–77 • (GAC)			

Date	Day	Assignment	Due Date	✓	Grade
		Second Semester-Fourth Quarter			
Week 1	Day 136	Read Devotional 1 • Pages 12–13 • *The Complete Zoo Adventure* (TCZA) • Complete Worksheet 3 • Page 137 • (TG)			
	Day 137	Read Introduction to Birds (lefthand side) and Flamingo Pages 31–32 • (TCZA) • Complete Worksheet 4 Page 138 • (TG)			
	Day 138	Read Peacock • Pages 33–35 • (TCZA) Complete Worksheet 5 • Page 139 • (TG)			
	Day 139	Read Hummingbird • Pages 36–38 • (TCZA) Complete Worksheet 6 • Page 140 • (TG)			
	Day 140	Read "Small, But Important" • Pages 78–79 • (GAC)			
Week 2	Day 141	Read Devotional 2 • Pages 14–15 • (TCZA) Complete Worksheet 7 • Page 141 • (TG)			
	Day 142	Read Parrot & Macaw • Pages 39–41 • (TCZA) • Complete Worksheet 8 • Page 142 • (TG)			
	Day 143	Read Eagles & Hawks • Pages 42–44 • (TCZA) • Complete Worksheet 9 • Page 143 • (TG)			
	Day 144	Read Owl • Pages 45–46 • (TCZA) • Complete Worksheet 10 Page 144 • (TG) (Using Biome Cards 141)			
	Day 145	Read "What A Dad!" • Pages 80–81 • (GAC)			
Week 3	Day 146	Read Devotional 3 • Pages 16–17 • (TCZA) Complete Worksheet 11 • Page 145 • (TG)			
	Day 147	Read Introduction to Mammals (lefthand side) and Fruit Bat • Pages 47–49 • (TCZA) • Complete Worksheet 12 Page 146 • (TG)			
	Day 148	Read Wolf • Pages 50–52 • (TCZA) Complete Worksheet 13 • Page 147 • (TG)			
	Day 149	Read Polar Bear • Pages 53–55 • (TCZA) Complete Worksheet 14 • Page 148 • (TG)			
	Day 150	Read "Armored Protection" • Pages 82–83 • (GAC)			
Week 4	Day 151	Read Devotional 4 • Pages 18–19 • (TCZA) Complete Worksheet 15 • Page 149 • (TG)			
	Day 152	Read Panda • Pages 56–57 • (TCZA) • Complete Worksheet 16 Page 150 • (TG)			
	Day 153	Read Koala • Pages 58–59 • (TCZA) • Complete Worksheet 17 Page 151 • (TG) (Using Biome Cards 141)			
	Day 154	Read Kangaroo • Pages 60–61 • (TCZA) • Complete Worksheet 18 • Page 152 • (TG)			
	Day 155	Read "Do Your Work Well" • Pages 84–85 • (GAC)			
Week 5	Day 156	Read Devotional 5 • Pages 20–21 • (TCZA) Complete Worksheet 19 • Page 153 • (TG)			
	Day 157	Read Lion • Pages 62–64 • (TCZA) • Complete Worksheet 20 Page 154 • (TG) (Using Biome Cards 141)			
	Day 158	Read Meerkat • Pages 65–66 • (TCZA) • Complete Worksheet 21 • Page 155 • (TG)			
	Day 159	Read Hyrax • Pages 67–68 • (TCZA) • Complete Worksheet 22 • Page 156 • (TG) (Using Biome Cards 141)			
	Day 160	Read "Love One Another" • Pages 86–87 • (GAC)			

Date	Day	Assignment	Due Date	✓	Grade
Week 6	Day 161	Read Devotional 6 • Pages 22–23 • (TCZA) Complete Worksheet 23 • Page 157 • (TG)			
	Day 162	Read Chimpanzee • Pages 69–71 • (TCZA) Complete Worksheet 24 • Page 159 • (TG)			
	Day 163	Read Gorilla • Pages 72–74 • (TCZA) Complete Worksheet 25 • Page 160 • (TG)			
	Day 164	Read Introduction to Mammals (lefthand side) and Giraffe Pages 75–77 • (TCZA) • Complete Worksheet 26 Page 161 • (TG)			
	Day 165	Read "I See You!" • Pages 88–89 • (GAC)			
Week 7	Day 166	Read Devotional 7 • Pages 24–25 • (TCZA) Complete Worksheet 27 • Page 163 • (TG)			
	Day 167	Read Zebras & Horses • Pages 78–80 • (TCZA) Complete Worksheet 28 • Page 165 • (TG)			
	Day 168	Read Camel • Pages 81–83 • (TCZA) Complete Worksheet 29 • Page 166 • (TG)			
	Day 169	Read Elephant • Pages 84–86 • (TCZA) Complete Worksheet 30 • Page 167 • (TG)			
	Day 170	Read "Generous Doctors" • Pages 90–91 • (GAC)			
Week 8	Day 171	Read Rhinoceros • Pages 87–89 • (TCZA) Complete Worksheet 31 • Page 168 • (TG)			
	Day 172	Read Hippopotamus • Pages 90–92 • (TCZA) Complete Worksheet 32 • Page 169 • (TG)			
	Day 173	Read Introduction to Reptiles (lefthand side) & Komodo Dragon • Pages 93–95 • (TCZA) Complete Worksheet 33 • Page 170 • (TG)			
	Day 174	Read Alligators & Crocodiles • Pages 96–98 • (TCZA) Complete Worksheet 34 • Page 171 • (TG)			
	Day 175	Read "No Vacancy" and "Deep Roots" • Pages 92–95 • (GAC)			
Week 9	Day 176	Read Tortoise • Pages 99–100 • (TCZA) • Complete Worksheet 35 • Page 172 • (TG)			
	Day 177	Read Introduction to Amphibians (lefthand side) & Tree Frogs • Pages 101–103 (TCZA) Complete Worksheet 36 • Page 173 • (TG)			
	Day 178	Complete Worksheet 37 • Page 174 • (TG)			
	Day 179	**Take Quiz 2 - Animal Observations** • Page 179 • (TG)			
	Day 180	Zoo Field Trip or Zoo in the House Day • Page 182 • (TG)			
		Final Grade			

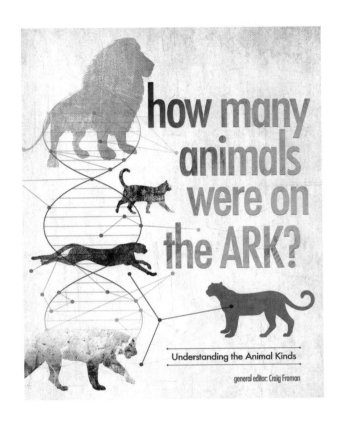

how many
animals
were on
the ARK?

Understanding the Animal Kinds

general editor: Craig Froman

Worksheets

for Use with

How Many Animals Were on the Ark?

Teacher Note: Some students will find the information in *How Many Animals Were On The Ark?* to be challenging. Students may need to read the material, work on their activity or worksheet, and then go back to the student book if needed to help complete the worksheet. Please take time to discuss the topics presented and go over unfamiliar vocabulary with the student.

Color the image of the Ark.

By faith Noah, being divinely warned of things not yet seen, moved with godly fear, prepared an ark for the saving of his household, by which he condemned the world and became heir of the righteousness which is according to faith.
—Hebrews 11:7

Fill in the blanks with the correct word from the Word Bank.

different   ten   features   living   kind

1. _____ times in Genesis 1 the phrase "according to its [or their] kind" is used in connection with different types of plants and animals.

2. Since two of each kind of land animal (and seven pairs of some) were brought aboard the Ark for the purpose of preserving their offspring upon the earth (Genesis 7:3), it seems clear that a "kind" represents the basic boundary of a _____ thing.

3. The offspring of a living thing is always the same kind as its parents, even though it may have different _____.

4. Diverse breeds of dogs can produce offspring with each other — indicating that all dogs are of the same _____.

5. Dogs will not interbreed with cats, however, since they are a _____ kind.

Write about or draw your favorite cat or dog!

Look closely at the following tracks from different kinds of dogs.

Dog

Wolf

Fox

1. What is the same about them?

2. What is different about them?

3. Now take some time to draw one or two of them.

4. If you can, go outside with your teacher and see if you can find any animals tracks on your own!

Unscramble the missing word to fill in the blank.

1. Creation researchers have found that "kind" is often at the level of "_____" in the modern way we classify animals. **(lamify)**

2. And God made the beast of the earth according to its kind, cattle according to its kind, and everything that creeps on the earth according to its kind. And God saw that it was _____. (Genesis 1:25) **(odog)**

3. God placed the potential for tremendous _____ within the original created kinds. **(vayrite)**

4. Baraminology is a word that comes from two _____ words: *bara*, meaning "created," and *min*, meaning "kind." **(Hewbre)**

5. The study of baraminology attempts to classify fossil and living organisms into their original created kinds (or _____). **(bamirnas)**

Draw a line to the correct missing word.

1. The familiar system of scientific

   _____ was developed by a

   creationist and biologist, Carl Linnaeus.

   Kingdom

2. In classification, the broader genus name is given

   first and capitalized, followed by the narrower

   _____ name in lower case, both in italics.

   naming

3. All flesh is not the same flesh, but there is one kind of flesh of

   men, another flesh of animals, another of fish, and another of

   _____.

   (1 Corinthians 15:39)

   birds

4. The part of biological classification that emphasizes giving

   proper scientific names to organisms is called "

   _____."

   species

5. Linnaeus also gave us a series of taxonomic ranks for grouping

   created kinds into a series of categories that you may be

   familiar with: family, order, class, phylum (or division in

   plants), and _____.

   taxonomy

See if your teacher can help you research your favorite animal's biological classification, just like the lion shown in your *How Many Animals* book!

Draw a line to the correct missing word.

1. Taxonomy is just a fancy word for _____ or

    classifying living things.   **Mammals**

2. A good classification system of animals and plants should start

    with the account of _____ in Genesis.   **family**

3. Linnaeus always spoke of the species as "the created kinds,"

    though creation scientists today know that the level of

    "_____" is a better approximation of   **tree**

    kind.

                                                                                 Creation

4. Evolutionary beliefs about classification are usually represented

    as a "_____ of life."

5. _____ are a group defined as animals   **sorting**

    that nourish their young on milk (from mammary glands),

    whether they run (horses), swim (whales), fly (bats), or burrow

    (moles).

Color the many different varieties of animals God created.

Fill in the blanks with the correct word from the Word Bank. Answer as many as you can.

dats   species   min   kind   dogs

1. A good rule of thumb is that if two things can breed together, then they are of the same created

    _____.

2. In the world today there are no reports of _____ (dog + cat) or hows (horse + cow).

3. Often, people are confused into thinking that a "_____" is a "kind."

4. The Bible's first use of this word "kind" (Hebrew: _____) is found in Genesis 1 when God creates plants and animals "according to their kinds."

5. When _____ breed together, you get dogs; so there is a dog kind.

Unscramble the missing word or number to fill in the blank.

1. In today's culture, many people have been led to believe that animals and plants have been like this for tens of thousands of years and perhaps _____ of years. **(imlilosn)**

2. From a biblical perspective, land animals like wolves, zebras, sheep, lions, and so on each have at least two ancestors that lived on Noah's Ark, only about _____ years ago. **(0,403)**

3. Animals have undergone many _____ since that time, but dogs are still part of the dog kind and cats are still part of the cat kind. **(cehangs)**

4. Baraminology is a field of study that attempts to classify fossils and living organisms into _____ or kinds. **(bamirnas)**

5. Baramin is commonly believed to be at the level of _____ and possibly order for some plants/animals. **(ayfmil)**

Sketch some of the landfowl feathers from page 17 in the *How Many Animals* book or feathers you may have found outside.

Fill in the blanks from your reading.

1. If two animals can produce a _____, then they are considered to be of the same kind.

2. Hybrids are animals that are created from mixing two different _____ of the same kind.

3. Bring out with you every living thing of all flesh that is with you: birds and cattle and every creeping thing that creeps on the earth, so that they may abound on the earth, and be fruitful and multiply on the earth. (_____ 8:17)

4. If kind is typically at the level of family/order, there would have been plenty of room on the Ark to take two of every kind and _____ pairs of some.

5. Even though many different dinosaurs have been identified, creation scientists think there are only about 60–_____ "kinds" of dinosaurs.

Color some of the different varieties of dinosaurs God created and draw one of your own.

Hadrosaurus

Germanodactylus

Inostrancevia

My Dinosaur

Use the code to find the missing words.

a	1		j	10		s	19
b	2		k	11		t	20
c	3		l	12		u	21
d	4		m	13		v	22
e	5		n	14		w	23
f	6		o	15		x	24
g	7		p	16		y	25
h	8		q	17		z	26
i	9		r	18			

___ ___ ___ ___
13   21   12   5

___ ___ ___ ___ ___
12   9   15   14   19

___ ___ ___ ___ ___ ___
12   9   7   5   18   19

___ ___ ___ ___ ___ ___ ___ ___
19   20   18   9   16   9   14   7

___ ___ ___ ___ ___ ___ ___ ___
23   8   15   12   16   8   9   14

Fill in the blanks with the correct "secret" word!

1. One of the most popular hybrids of the past has been the _____, the mating of a horse and donkey.

2. Zonkeys and zorses have a mixture of their parents' traits, including the beautiful _____ patterns of the zebra parents.

3. The _____ is the result of mating a false killer whale and bottlenose dolphin.

4. _____ are the largest cats in the world, weighing in at over 1,000 pounds.

5. Tigons are the result of mating a male tiger and a female _____.

Fill in the blanks from your reading.

1. It made sense to Carl Linnaeus to begin using _____ terms for his new

    classification system.

2. After Linnaeus, _____ and species were commonly used in modern biological

    classification systems with slightly different definitions.

3. The word _____ was viewed for a long time as the biblical "kind."

4. The idea of one kind of animal changing into another has never been _____.

5. Great variety can be observed in the offspring of animals of the same kind, just as the same cake

    _____ can be used to make many different cakes.

Unscramble the missing word or number to fill in the blank.

1. Centuries ago, English breeders saw a need for a massive _____ that could guard, do

    search and rescue, and assist police work, and so the English mastiff was born. **(gdo)**

2. There are about 500 _____ dog breeds around the world. **(doticmes)**

3. Does it amaze you that the vast majority of dog breeds are fewer than _____ years old?
    (050)

4. For since the creation of the world His invisible attributes are clearly seen, being understood by

    the things that are made, even His eternal power and Godhead, so that they are without excuse.

    (_____ 1:20) **(ansRmo)**

5. Despite loving his pets dearly and constantly marveling at their personalities and abilities,

    _____ was unwilling to credit their origin with the Creator. **(inwarD)**

Do the "Animals Kinds" word search.

DINOSAUR	BEAGLE	ELEPHANT	CANARY
HORSE	TIGER	GOLDFISH	RABBIT

```
C  E  L  E  P  H  A  N  T  E
A  M  D  I  N  O  S  A  U  R
N  F  B  Z  U  R  Z  O  S  B
A  P  R  I  N  S  S  D  O  E
R  Y  I  U  T  E  W  Y  B  A
Y  P  N  K  A  U  R  I  E  G
E  S  G  T  I  G  E  R  A  L
R  A  B  B  I  T  L  V  G  E
P  A  P  E  R  V  C  X  L  D
G  O  L  D  F  I  S  H  E  S
```

Fill in the blanks from your reading.

1. Darwin and the scientific community of his time had no understanding of the complex

    _____ required to produce variation.

2. Dogs are the focus of ongoing research to solve the mystery of vast variations within a kind over a

    _____ amount of time.

3. The study of domesticating or _____ animals has been broken down into two

    major categories: biological changes and behavioral changes.

4. As dogs became more domesticated, many came to depend on _____ for

    survival, no longer using their hunting instincts or actually bred to be gentle family pets.

5. "Dog" artifacts do not appear until near the end of the Ice Age, about _____ years ago.

Sketch an image of this fierce chihuahua!

Fill in the blanks with the correct word or number from the Word Bank.

variety   multiply   orchid   **Babel**   **127**

1. And God blessed them, saying, "Be fruitful and multiply, and fill the waters in the seas, and let birds _____ on the earth." (Genesis 1:22)

2. We find at least _____ post-Flood species in the dog/wolf family alone, which arose from the first parents on the Ark.

3. We still do not have sufficient evidence to settle whether the human population at _____ had already begun domesticating animals from the Canidae family.

4. The _____ family is probably the largest flower family in the world, with thousands of wild species filling the earth since Noah's Flood.

5. By God's gracious design, artificial selection allows us to reveal _____ never before seen.

Draw a line to the correct missing word.

1. _____ are more varied than any other animal, whether domestic or wild.

   obedience

2. Thousands of years would not be considered a long time to evolutionists; in fact, many consider it
"_____."

   quick

3. Because evolutionary scientists are trained to leave the Creator out of the equation, they assume dogs needed tens of thousands of years of natural selection acting on random mutations to produce _____.

   Dogs

4. A German shepherd's fur grows in a _____ layer that gives them great protection against the elements, perfect for guard dogs in cold environments.

   changes

5. German shepherds also have a temperament that includes courage, strength, loyalty, focus, and _____.

   double

Color the different types of dogs God created.

Fox

Wolf

German shepherd

French bulldog

Fill in the blanks from your reading.

1. The pug was developed over 2,000 years ago in _____.

2. Traits of _____ dogs include a great ability to control the movement of other animals and deep loyalty.

3. Traits of _____ dogs include a high intelligence, strength, and solid companionship.

4. Traits of _____ dogs include strength, intelligence, and fierce loyalty to watch over property.

5. Traits of _____ dogs include active metabolism, alert nature, and remarkable instincts in the water and woods.

Fill in the blanks from your reading.

1. National Geographic aired a program that detailed information about the first documented grizzly/ polar bear hybrid (_____ bear) in the wild.

2. The animal from "Mystery Bear of the Arctic" was mostly _____ with brown splotches but had the grizzly traits of long claws, concave face, and humped back.

3. The word bear finds its origins in the Indo-European root _____.

4. People have cherished the likeness of bears as toys, political emblems, and symbols of strength and great _____.

5. The survival of many bear species is a conservation concern, which is why they are probably the most studied _____.

Use the code to find the missing words.

a	1		j	10		s	19
b	2		k	11		t	20
c	3		l	12		u	21
d	4		m	13		v	22
e	5		n	14		w	23
f	6		o	15		x	24
g	7		p	16		y	25
h	8		q	17		z	26
i	9		r	18			

___ ___ ___ ___
14 15 1   8

___ ___ ___ ___ ___
1   4   1   16 20

___ ___ ___ ___
6   15 15 4

___ ___ ___ ___ ___ ___ ___ ___
4   5   19 9   6   14 5   4

___ ___ ___ ___ ___ ___ ___ ___
3   18 5   1   20 9   15 14

Fill in the blanks with the correct "secret" word!

1. After _____ released the animals, the world they reentered was not the same.

2. It was a world where God's creatures had to be able to _____ to new and drastically changing conditions.

3. Bears are known for their adaptability to new environments and _____ sources.

4. Worldview is critical to how we interpret the _____ around us.

5. This issue is about a struggle between a biblical worldview that states that life is _____ versus a humanistic worldview that states that life is random

Take time to sketch the different bear tracks.

The Bible is very clear that people were created differently from animals. And not just differently, but actually in God's image and likeness, and placed in dominion or authority over the creatures, to care for them as good leaders do.

> *Then God said, "Let Us make man in Our image, according to Our likeness; let them have dominion over the fish of the sea, over the birds of the air, and over the cattle, over all the earth and over every creeping thing that creeps on the earth."* —Genesis 1:26

The reading today talked about people and the variation or differences that appear in us. There are some who might not think it possible that our two ancestors (Adam and Eve) could fill the world with people who have such different colored hair and skin. But we see that it could be done, and in less than 100 years!

The book of Acts (17:26) in the Bible states that God "…made from one blood every nation of men to dwell on all the face of the earth…" So there aren't many "races" of people, but only one race, and we all came from the same two ancestors, Adam and Eve. We are truly one family of people. Think about the following questions and write your answers.

1. What does melanin do in our skin?

2. What is one reason it's important to know that we are all one race and not from many races?

3. How might it help people live more at peace knowing we are all a part of the same family?

Draw a line to the correct missing word.

1. Great _____ in size, color, form, function, etc., would also be present in the created ancestors of all the other kinds.

   reproduce

2. Varieties within a created kind have the same genes, but different alleles (or versions of the _____).

   blood

3. In many Native American tribes, there is a high percentage of _____ type A, but that type is quite rare among other tribes.

   variation

4. All the different varieties of human beings can marry one another and have _____.

   gene

5. Many varieties of plants and animals retain the ability to _____, despite differences in appearance as great as those between St. Bernards and Chihuahuas.

   children

Fill in the blanks from your reading.

1. There would have been no more than 16,000 land animals and birds on the Ark, though more likely closer to _____.

2. According to the Bible, the Ark had _____ decks (floors).

3. God would likely have sent to Noah _____ (and therefore small, but not newborn) representatives of the larger animals.

4. The animals on the Ark could have survived their year-long journey as _____.

5. _____ preserved through means of drying, pickling, salting, or smoking could have been provided.

Color these scenes from inside the Ark.

Unscramble the missing word to fill in the blank.

1. There is a big difference between the long-term care required for animals kept in _____ and the temporary, emergency care required on the Ark. (**sooz**)

2. Drinking water could have been piped into _____, just as the Chinese have used bamboo pipes for this purpose for thousands of years. (**trghsou**)

3. As much as 6 U.S. _____ of animal waste may have been produced daily on the Ark! (**sont**)

4. The danger of toxic or explosive manure gases would be alleviated by the constant _____ of the Ark. (**momentve**)

5. Another possibility on the Ark is that the animals may have _____ during a portion of the voyage. (**berhitedna**)

Fill in the blanks from your reading.

1. It is important to understand that the word "_____" used in Genesis 1 seems to represent something closer to the "family" level of classification in most instances.

2. The actual number of mammals on the Ark could easily have been well over _____.

3. Analysts conducting research for the Ark Encounter project predict that less than _____ animal kinds were preserved on Noah's Ark.

4. Other than the _____, the Ark is the greatest reminder of salvation, for in the judgment God provided salvation to Noah and his family.

5. Jesus said, "I am the _____, by me if any man enters in he will be saved."

Fill in the blanks with the correct word from the Word Bank.

rhinos   plates   tusks   Chalicotheres   spikes

1. There were probably under 1,400 animal _____ on the Ark.

2. Cynognathids were actually more like _____ than crocodiles or lizards.

3. Spinosaurs were large predatory dinosaurs known for their long, crocodile-like heads, massive hooked claws, and _____-backs.

4. Once considered the relatives of _____, entelodonts were a unique and bizarre animal kind.

5. Long considered "man's best friend," not every member of the _____ family is like our beloved pets.

Measuring Activity. Get some yarn or string, a measuring tape, a pair of scissors, and the following animal sizes chart to help you see just how big or small certain animals are! Have your teacher help you tie a knot approximately every foot on a 10-foot long piece of yarn or string, using the scissors to cut the string or yarn you are using. Then stretch the long string or yarn across the floor to see how you might compare to the animal heights or lengths.

1.  Giraffe — 19 feet

2.  Elephant — 12 feet

3.  Ostrich — 8 feet

4.  Horse — 7 feet

5.  Llama — 6 feet

6.  Gorilla — 5 feet

7.  Pig — 4 feet

8.  Dog — 3 feet

9.  Cat — 1 foot

10.  Mouse — 4 inches

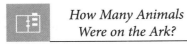
Fill in the blanks from your reading.

1. The trademark thick woolly coat of domestic sheep is not a _____ trait.

2. _____ are a very diverse group of animals that include buffalo, bison, and certain antelopes.

3. Pachycephalosaurs weren't _____ with their thick skull caps.

4. Hippo species have come in various sizes and slightly different shapes, but it seems they really haven't changed much overall since the _____.

5. Popularly touted as _____ ancestors, pakicetids were a small family of land-dwelling or semi-aquatic mammals not represented in the world today.

Fill in the blanks with the correct word from the Word Bank.

| four | okapis | beak | headgear | sauropod |

1. The stahleckeriid had a stocky body, short tail, _____, and tusk-like facial flanges.

2. As far as we know, certain _____ dinosaurs were the largest animals that ever walked the earth.

3. Parasaurolophus bore some of the most striking _____, which they probably used as vocal resonators.

4. Compared with elephants, palaeomastodonts were smaller, had shorter trunks, and grew _____ tusks rather than two.

5. Living representatives of the giraffe kind include giraffes and _____.

Fill in the blanks from your reading.

1. It seems pigs haven't changed much in appearance in the last _____ years or so.

2. The earliest-known _____ were more like smaller varieties from South America — guanacos, llamas, vicuñas, and alpacas.

3. Today, the cat kind is represented by felines, like cheetahs, and pantherines, like _____.

4. It's good to remember that not every _____ animal that looked like a dinosaur actually was a dinosaur.

5. Simosuchus was a pug-nosed, short-tailed, _____-like reptile.

Color the many different varieties of Ark creatures God created.

Unscramble the missing word to fill in the blank.

1. Fossil evidence confirms the _____ of the stegosaurus were used as weapons.

    (ikesp)

2. The _____ of the stegosaurus were likely display and body heat–regulating organs.

    (spatle)

3. Uintatheres probably looked a lot like _____, though these two groups are completely unrelated. **(osinrh)**

4. Many uintatheres had _____, ossicones like giraffes, and elephant-like feet. **(sktus)**

5. _____ were a group of odd-toed ungulates — like horses, rhinos, and tapirs. **(eresChalcothi)**

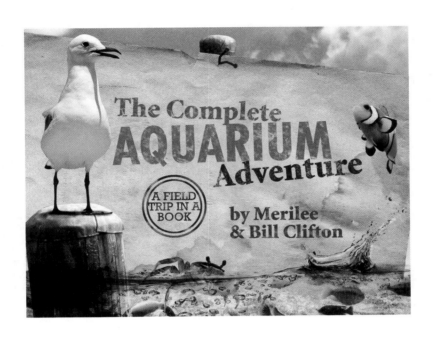

Worksheets

for Use with

The Complete Aquarium Adventure

Creation

All CREATION points us to God and reminds us that He is all-knowing and all-powerful! He made the universe, the earth, and all that is in it in six 24-hour days. Can you "create" at least 24 words from the letters in CREATION?

Instructions: Use each letter only once. Words must contain three or more letters. No proper names.

C R E A T I O N

Find and Color — Birds

Find and color the following birds: anhinga (*Anhinga anhinga*); double-crested cormorant (*Phalacrocorax auritus*); brown pelican (*Pelecanus occidentalis*); and African penguin (*Spheniscus demersus*).

For thou, Lord, hast made me glad through thy work: I will triumph in the works of thy hands.
—Psalm 92:4 (KJV)

Who's Who?

Can you identify the Anhinga and the Double-crested Cormorant in the following images? Label them correctly.

1. _____   2. _____

3. How could you tell the difference between these birds?

4. List three things these two birds have in common?

Connect the Dots — Birds

First catch of the day!

Draw in the watery environment that most often surrounds this amazing bird, the brown pelican (*Pelecanus occidentalis*), as well as its most recent catch.

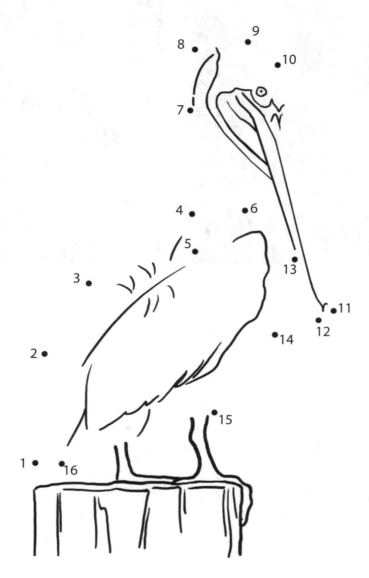

God created birds on day _____.

Water Cycle Bracelet

All the rivers run into the sea; yet the sea is not full; unto the place from whence the rivers come, thither they return again. —Ecclesiastes 1:7 (KJV)

This verse describes the water cycle that God put in motion to provide mankind with an unending source of life-giving water. This water moves from the earth to the atmosphere and back to the earth again and again and again. The water cycle bracelet reminds us of God's plan and care for us.

Supplies Needed

- One pony bead of yellow, blue, green, white, clear, and brown as listed on page 148 of *The Complete Aquarium Adventure* book to represent different parts of the water cycle. Optional: You may fill the entire bracelet using 4–5 beads of each color (depending on wrist size) to represent the unending cycle.

- Elastic cord cut to 8 inches (or you may also use any kind of string, twine, yarn, etc.)

Bracelet Directions

In the order listed below, thread the colored beads onto the elastic cord. (Hint: Tape one end of the cord to a solid surface so beads won't slip off while you are stringing them.) Repeat as desired with more sets of colored beads in the order listed. Put the two ends of the cord together and tie a knot. Trim excess cord. Wear around your wrist. Use the bracelet to explain God's plan for recycling water to a family member or friend.

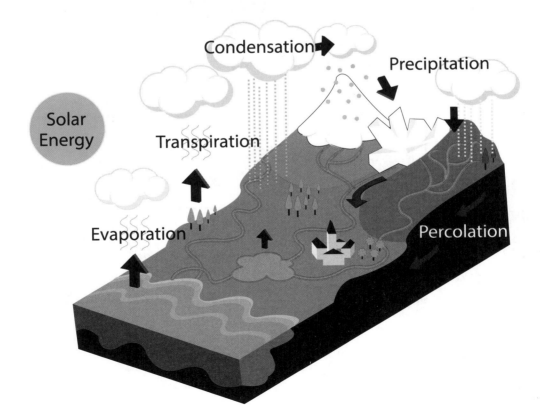

Turtle Tracks

Help the baby loggerhead sea turtle find its way from its nest, across the sand, and into the sea to the floating sargassum weed where it will feed and grow. Be sure to avoid dangers like gulls, raccoons, dogs, and houses with lights!

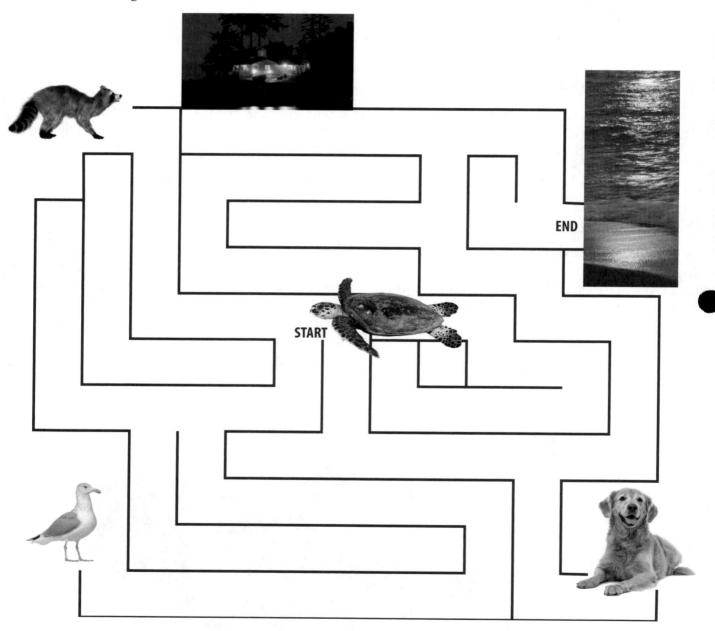

Find and Color — Fish

Find and color the following fish: common clownfish (*Amphiprion ocellaris*); seahorse (*Hippocampus*); porcupinefish (*Diodon holocanthus*); lionfish (*Pterois volitans*); green moray eel (*Gymnothorax funebris*); bonnethead shark (*Sphyrna tiburo*); cownose ray (*Rhinoptera bonasus*).

He hath made his wonderful works to be remembered: the LORD is gracious and full of compassion.
—Psalm 111:4 (KJV)

Color by Number

Add or subtract numbers to find the correct color to use in each space. Then color the clownfish, a common aquarium fish.

8 = Green   5 = Orange   11 = Blue

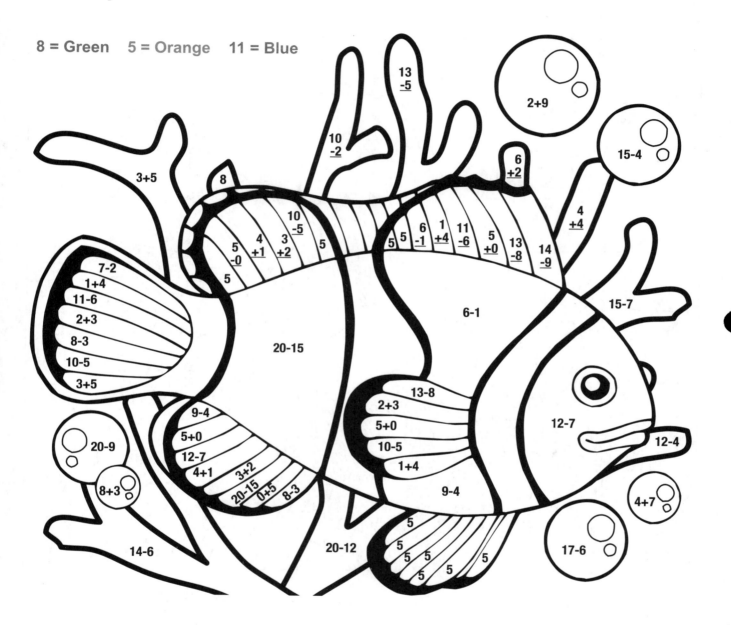

Guess Who? (Game)

Using the "Animal Fact Cards" from the Tool Kit and paper clips or clothespins, attach a card to each student's shirt near the neck in the back, making sure the student does not see the animal. When leader says, "Begin," students ask each other questions that can be answered only with "Yes" or "No." Examples: "Do I have fins?" "Do I breathe air?" "Do I live in water?" "Do I have fur?" Ask only one question per student and then go to another student to ask another question. The first student to identify his or her animal wins!

I meditate on all thy works; I muse on the work of thy hands.

—Psalm 143:5 {KJV)

The Great Barracuda

Answer the questions about barracudas below.

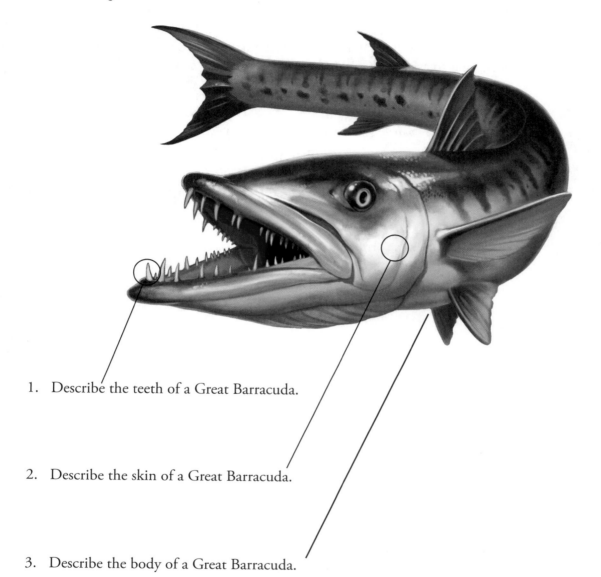

1. Describe the teeth of a Great Barracuda.

2. Describe the skin of a Great Barracuda.

3. Describe the body of a Great Barracuda.

4. How did the way a Great Barracuda eats change after the Fall of Man?

Decoding

Write the letter of the alphabet or number that matches the nautical flag in the spaces below to decode the puzzle.

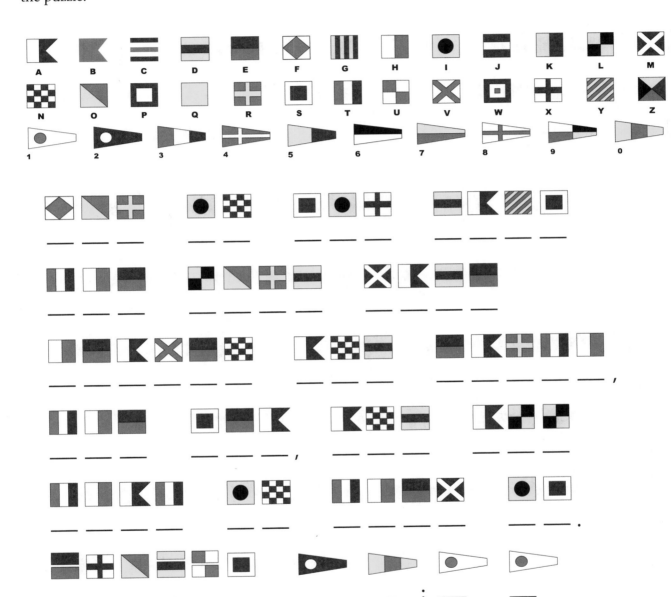

Connect the Dots — Mammals

Out here the water is cold. Illustrate the icy blue environment of this beluga whale (*Delphinapterus leucas*) as it searches for fish to eat.

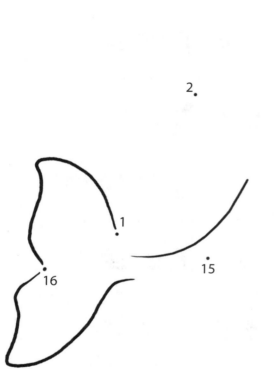

Sea Hunt 1

M	K	L	A	U	P	S	I	D	E	D	O	W	N	J	E	L	L	Y	D	U	W	Z	K
K	L	G	L	H	G	P	H	A	R	B	O	R	S	E	A	L	B	N	Q	S	J	B	I
M	O	S	L	P	B	K	P	F	L	Y	C	R	T	J	P	J	Z	W	R	W	O	F	L
I	C	E	I	E	E	W	A	W	U	K	D	O	N	O	C	N	G	N	A	N	Z	C	L
D	E	A	G	L	L	P	C	L	B	K	J	H	R	S	E	A	T	U	R	T	L	E	E
C	A	A	A	E	U	G	I	M	X	J	L	U	K	A	P	R	W	G	Y	H	D	P	R
A	N	N	T	W	G	G	F	S	V	W	H	A	L	E	L	F	T	W	Z	C	Q	V	W
W	D	E	O	M	A	O	I	J	N	X	V	O	L	O	G	G	E	R	H	E	A	D	H
B	J	M	R	R	G	S	C	A	M	V	C	D	R	I	K	F	U	L	C	D	H	H	A
F	K	O	D	O	L	P	H	I	N	H	O	R	S	E	S	H	O	E	C	R	A	B	L
K	I	N	I	Y	F	S	U	R	Z	T	B	C	E	B	O	T	T	L	E	N	O	S	E
W	A	E	W	B	E	H	E	S	V	V	O	F	A	Z	U	H	G	H	D	U	N	N	S
K	P	U	U	Y	D	V	Q	Q	O	N	H	C	S	P	B	X	X	R	O	G	P	Q	W
X	Y	Y	C	Z	O	O	Y	S	Y	X	R	M	T	D	L	G	U	P	O	G	V	U	L
T	F	B	M	O	O	N	J	E	L	L	Y	X	A	O	W	B	S	E	A	L	I	O	N
M	J	W	Q	N	M	K	U	A	L	Z	R	T	R	F	P	B	B	S	M	V	X	S	Y
G	P	A	T	L	A	N	T	I	C	H	L	Y	W	A	L	U	G	M	P	T	W	F	U
F	V	W	S	C	N	X	O	B	E	N	C	H	F	F	R	Z	S	Y	U	V	F	D	W

Coral	Sea star	Whale	Sea turtle
Giant octopus	Upside-down jelly	Sea lion	Loggerhead
Horseshoe crab	Bottlenose	Harbor seal	Atlantic
Moon jelly	Dolphin	Killer whale	Pacific
Sea anemone	Beluga	Alligator	Ocean

| The Complete Aquarium Adventure | Read Lionfish Pages 53–54 | Day 62 | Worksheet 14 | Name |

By the Numbers!

Can you remember these fascinating facts about lionfish? Circle the correct answer.

1. They have been found at depths of _____ feet.

    2,000   200   20

2. They can grow to be _____ inches.

    15   20   12

3. They weigh between _____pounds.

    3 to 4   2 to 3   1 to 2

4. What day of the Creation week were lionfish created?

    Day 4   Day 5

What's in a Name

5. Can you list three of the four other names that lionfish are known for?

a.

b.

c.

Bonus Question!

What do lionfish use their "feathery" fins for?

Color the porcupinefish.

Color the seahorse.

Sea Hunt 2

B	Y	R	S	S	B	C	G	F	X	E	M	A	R	I	N	E	D	X	K	C	C	N	I
H	X	P	H	A	A	S	L	S	U	V	Y	N	U	B	B	Y	K	N	Z	R	B	J	M
F	P	K	A	N	R	P	B	O	O	N	L	R	V	T	C	Y	M	V	Z	E	H	S	O
P	K	N	R	D	R	M	O	Z	W	U	J	T	K	E	V	O	H	N	Y	R	J	M	R
P	I	M	K	T	A	H	C	R	J	N	T	Q	E	B	U	F	R	J	C	U	S	W	A
J	M	J	E	I	C	E	S	N	C	N	F	H	Q	Z	O	T	Q	M	Z	L	C	Z	Y
B	Q	K	D	G	U	X	B	E	N	U	U	I	E	C	Q	N	H	T	O	H	G	D	E
I	R	I	O	E	D	X	E	D	A	U	P	N	S	R	O	J	N	R	M	R	H	Q	E
B	F	O	J	R	A	A	U	S	P	H	R	I	F	H	N	W	O	E	Y	G	A	Z	L
O	Y	T	W	K	R	L	J	U	E	R	O	S	N	X	P	S	N	W	T	Q	F	N	N
F	D	D	Z	N	M	P	I	P	T	A	E	R	E	E	O	W	T	O	W	H	U	C	T
J	I	G	D	B	P	W	A	O	R	F	D	D	S	S	F	C	S	I	S	P	E	C	Z
U	P	F	G	A	W	E	N	N	N	U	C	R	A	E	H	I	E	L	N	E	R	A	Z
G	G	U	S	T	T	N	L	C	H	F	P	U	A	T	Y	A	S	A	P	G	R	E	D
Y	D	S	P	M	T	E	O	I	U	I	I	A	L	G	O	J	R	H	N	Z	R	A	Y
A	I	K	G	T	A	Y	Z	W	C	S	N	S	Q	E	O	R	R	K	E	P	M	A	Y
C	M	J	U	P	O	H	N	Z	Y	A	S	G	H	Z	K	N	G	H	Z	A	K	T	Y
S	K	P	O	N	U	N	V	U	T	G	N	S	A	D	Z	R	P	Y	Q	O	P	E	R

Anhinga	Cormorant	Nurse shark	Sea dragon
Barracuda	Cownose ray	Ocean	Seahorse
Bonnethead	Lionfish	Porcupinefish	Sand tiger
Brown pelican	Marine	Predator	Shark
Clownfish	Moray eel	Prey	Southern stingray

Connect the Dots — Bluespotted ribbontail ray

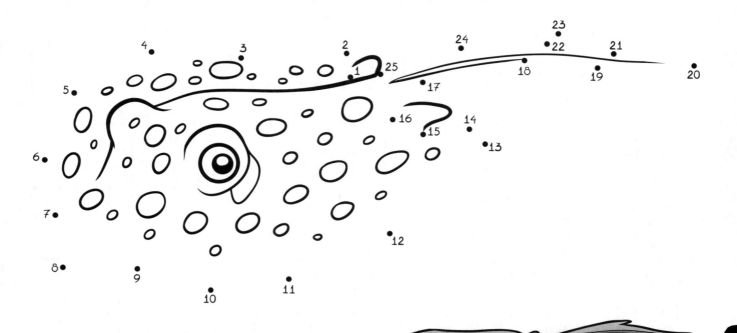

Color this undersea scene.

Color the nurse sharks below.

Top Marine Predator Mobile

Choose a top marine predator. This predator cannot be a baleen whale, herbivore, human, or extinct. Carefully research this animal to make sure that you can find enough information to replicate its food chain to three levels, ending with green plants. Get approval for your top marine predator choice from your teacher or parent before you begin work on your project.

Build a three-dimensional model of your top marine predator. Your model can be constructed of Styrofoam™, papier-mâché, balsa wood, fabric, or any other material that is light enough to be suspended from the ceiling and sturdy enough to suspend its food chain under it.

Using mobile construction, depict an accurate food chain of your top marine predator. Organisms suspended under the top predator may be two-dimensional (flat). As with a living food chain, your mobile should "move" if one part of the food chain is eliminated, leaving the chain out of balance. Do not go beyond hanging three organisms below any one animal, because the mobile will be too large to manage. For a top predator that can feed on almost anything, select three of the foods it prefers and suspend them under the animal. Be sure your mobile ends at the bottom of the food chain with green plants.

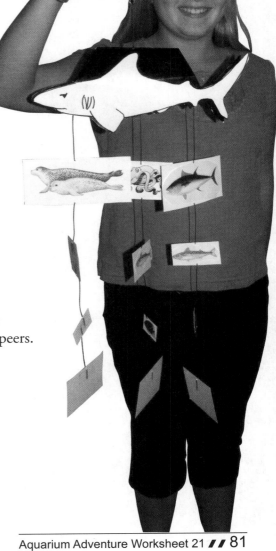

If this project is graded, the evaluation can be based upon any or all of the criteria listed below:

Mobile (100 points):

Creativity (25 points) — how top predator is made (three-dimensional), how levels below are displayed.

Accuracy (25 points) — shape/form of top predator, levels below pictured clearly, understandable.

Suspended properly (25 points) — from ceiling, "mobile" construction, balanced.

Craftsmanship (25 points) — overall look, finished details.

Oral Presentation (100 points): Student will research a top marine predator, then give an oral presentation to the student's family and peers.

Connect the Dots — Fish

Southern stingrays (*Dasyatis americana*) use their "wings" to soar through the water. Sketch in some of the rocky environment where it makes a home for itself.

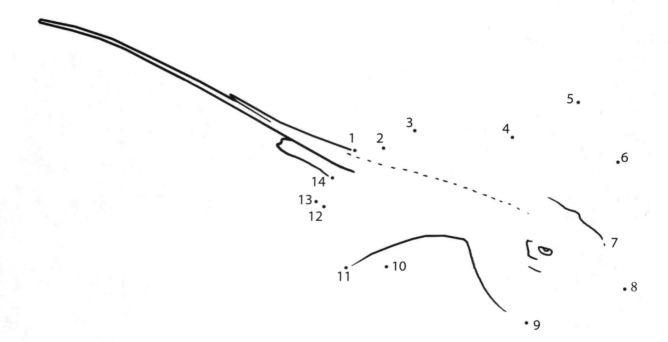

Fish Scramble

Unscramble the names of these fish. Write one letter on each line.

S O E S H A R E

T N A I S G R Y

I S O N F I L H

N E E B A D T O N H   K A S R H

Y L A F E   E S A   O R G A N D

W F S I H C O L N

A C R A B U R D A

S U N E R   K H A S R

N E G E R   Y M A R O   L E E

N A D S   G E T R I   K A S R H

The LORD on high is mightier than the noise of many waters, yea, than the mighty waves of the sea.
—Psalm 93:4 (KJV)

Aquarium Diorama

This project allows you to simulate a habitat that is suitable for marine creatures. Before you begin, decide which animals you would like to include in your habitat. Do some research to learn things such as where your animal lives and what it eats so you will know what is required for a suitable home.

Diorama directions:

1. Cut sections out of the long front and two short sides of a box (shoe box size or larger) so that it resembles a fish tank. (Younger children may need an adult's help.)

2. You can draw or glue pictures on the back wall for a background.

3. Tape plastic wrap inside the front and two side openings to resemble water.

4. Using construction paper or pictures, cut out fish and other sea animals that live in your habitat. Swimming animals may be suspended from the box top with thread.

5. To complete your diorama, make and add things that fit into your habitat such as sand, sea plants, rocks, bottom-dwelling animals, and coral reefs.

6. You may also want to prepare a short report to describe your project.

Then He said to them, "Follow Me, and I will make you fishers of men."
—Matthew 4:19

OOPS (Out of Place Sea) Animals

Cross out the sea creature that is different from the others in each row.

Seahorse	Leafy sea dragon	Sea anemone	Lionfish
Bottlenose dolphin	Bonnethead shark	Nurse shark	Sand tiger shark
Giant Pacific octopus	Cownose ray	Coral	Upside-down jellyfish
Killer whale	California sea lion	Harbor seal	Penguin

Find and Color — Invertebrates

Find and color the following invertebrates: jellyfish; upside-down jellyfish (*Cassiopeia xamachana*); sea anemone; coral; giant Pacific octopus (*Enteroctopus dofleini*); horseshoe crab (*Limulus polyphemus*).

But God hath chosen the foolish things of the world to confound the wise;
and God hath chosen the weak things of the world to confound the things which are mighty.

—1 Corinthians 1:27 (KJV)

Silhouette Identification

Identify the marine animals below and write each one's name (one letter per box) underneath the animal.

Because thou hast been my help, therefore in the shadow of thy wings will I rejoice.
—Psalm 63:7 (KJV)

The Complete
Aquarium Adventure

Read Sea Anemone ...
Pages 80–84

Day 79

Worksheet 28

Name

Connect the Dots — Invertebrates

This sea star has been looking for something tasty. Using a pencil or pen, design an oyster bed that it can feast upon.

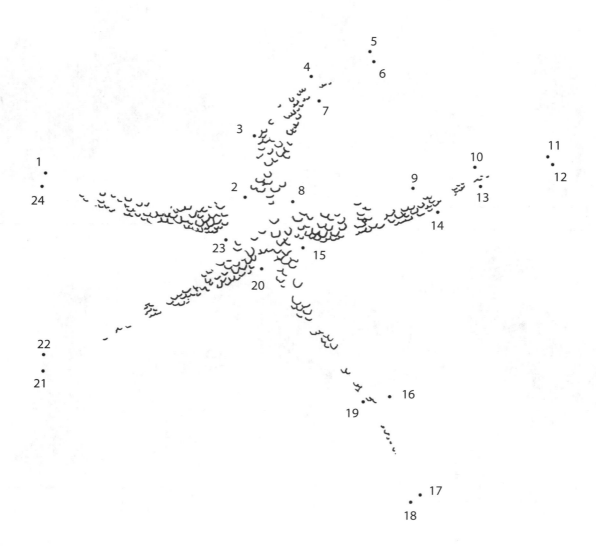

Bonus Activity

Do the sea star puzzle on page 183.

Graphic Artist — Beluga Whale

Copy the beluga whale from the left picture onto the graph paper on the right, square by square.

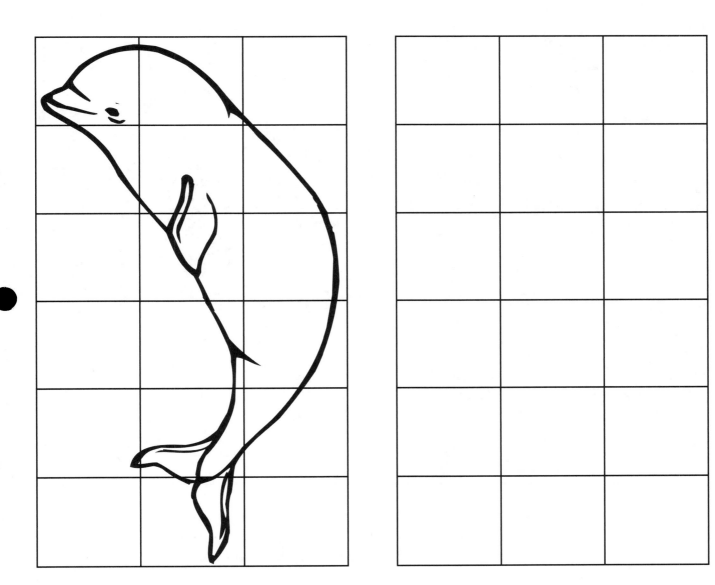

The beluga does not have a dorsal fin like other whales.
This design feature allows it to surface underneath the ice to breathe.

Find and Color — Mammals

Find and color the following mammals: bottlenose dolphin (*Tursiops truncatus*); killer whale (*Orcinus orca*); harbor seal (*Phoca vitulina*); California sea lion (*Zalophus californianus*).

And God created great whales, and every living creature that moveth, which the waters brought forth abundantly, after their kind.

—Genesis 1:21 (KJV)

Bonus Activity

Do the killer whale puzzle on page 185.

Survival Design Features

For each animal, describe your favorite survival design feature.

Anhinga	
Cormorant	
Pelican	
Penguin	
Clownfish	
Barracuda	
Green moray eel	
Leafy sea dragon	
Lionfish	
Porcupine fish	
Seahorse	
Bonnethead shark	
Nurse shark	
Sand tiger shark	
Cownose ray	
Southern stingray	
Coral	
Giant pacific octopus	
Horseshoe crab	
Moon jelly	
Upside-down jellyfish	
Sea anemone	
Sea star	
Dolphin	
Killer whale	
Beluga whale	
Harbor seal	
Sea lion	
Green sea turtle	
Loggerhead turtle	
American alligator	

Venn Diagram of Seal and Sea Lion

In the left circle, write characteristics of the seal. In the right circle, write characteristics of the sea lion. In the middle where circles overlap, write characteristics that both the seal and the sea lion share.

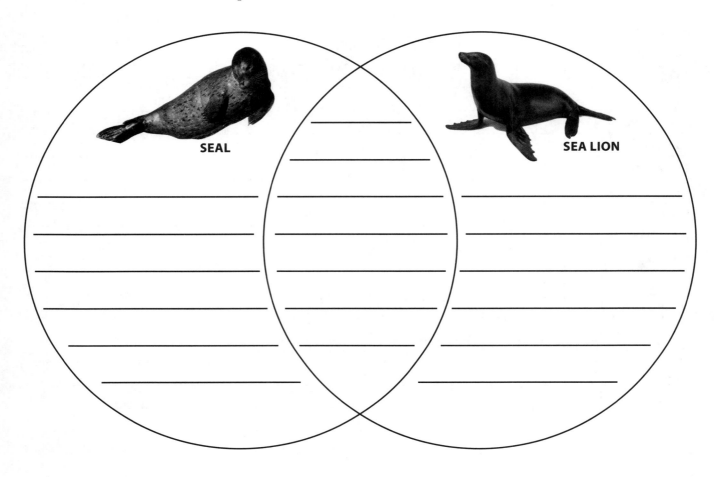

Spotted fur	Have whiskers (vibrissae)	No external ears
Solid color fur	Give birth on land	External ear flaps
Swim and feed in ocean	Round body	Four large flippers
Females smaller than males	Streamlined body	Four short flippers

Graphic Artist — American Alligator

Copy the alligator from the top picture onto the graph paper on the bottom, square by square.

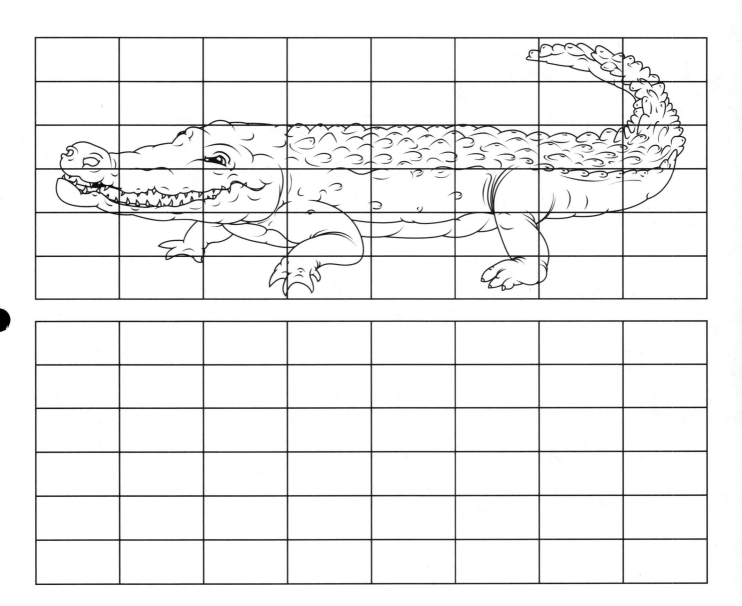

Baby alligators are dark green with yellow stripes to help camouflage them while they are most vulnerable.

Find and Color — Reptiles

Find and color the following reptiles: green sea turtle (*Chelonia mydas*); loggerhead sea turtle (*Caretta caretta*); American alligator (*Alligator mississippiensis*).

*The flowers appear on the earth; the time of the singing of birds is come,
and the voice of the turtle is heard in our land.*

—Song of Solomon 2:12 (KJV)

Bonus Activity! Enjoy the Green Sea Turtle Hand Puppet activity on page 181.

Connect the Dots — Reptiles

Baby alligators just hatched! Draw them in along with their nest and the scutes on the back of the mother American alligator (*Alligator mississippiensis*).

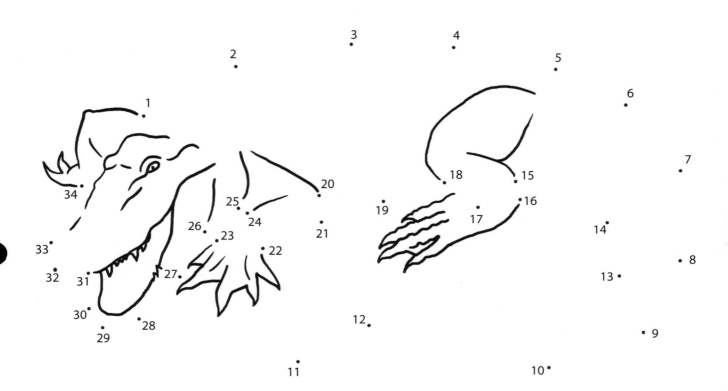

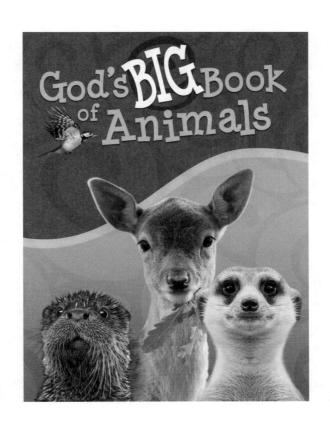

Worksheets

for Use with

God's Big Book of Animals

Fill in the blanks from your reading.

1. A group of hummingbirds is called a _____ of hummingbirds.

2. A hummingbird's heart can beat up to _____ times per minute.

3. Look at the birds of the air, for they neither sow nor reap nor gather into barns; yet your heavenly Father feeds them. Are _____ not of more value than they? (Matthew 6:26)

4. At night, hummingbirds slip into a kind of hibernation called "_____."

5. Since hummingbirds can _____, they don't need to slow down when they feed.

6. A group of toucans is technically called a _____ of toucans, but most people refer to them as a flock.

7. There is a _____ named after the bird called the Tucana.

8. Toucans have _____ wings, so they can only fly very short distances.

9. God created toucans with the ability to turn their heads completely _____ and rest their bills on their backs, fold their tails, and cover their heads.

10. Toucans eat fruit dipped in _____ and bird eggs.

Fill in the blanks from your reading.

1. The noise woodpeckers make when they peck trees is called "_____," and it is also a way they communicate with each other.

2. Woodpeckers make "cracking _____," which are areas they use for cracking nuts they find.

3. Woodpeckers do not have _____, so they swallow little grain-sized rocks with their meals, and that helps them grind their food.

4. A group of woodpeckers is called a _____ of woodpeckers.

5. Woodpeckers like to eat insects, especially _____ such as beetles and moths, as well as fruit, nuts, and berries.

6. A group of crows is called a _____ of crows.

7. Crows watch out for cats, eagles, or foxes, but their greatest fear is _____.

8. Studies show that a family of crows can eat about _____ insects at one meal.

9. Scientists have found that crows have more than _____ different calls they use to communicate.

10. If a crow is _____, his fellow crows will stay by his side, even in the face of danger.

Crossword Puzzle (Vultures and Owls)

Across

2. Their nests are built on high elevations to provide the most protection for their _____.

3. Owls cannot digest the fur and bones in their food so they usually cough those back up as something called a _____.

5. Unlike other birds, which hunt live prey, vultures feed on dead animals (_____).

7. God created owls with twice as many _____ bones as a human has so owls can rotate their heads in every direction.

8. Baby vultures feed on _____ crushed by their parents, which is how they get calcium.

9. In general, vultures are called a committee of vultures or a colony of vultures, unless they are circling in the air, when they are called a _____ of vultures.

Down

1. A group of owls is called a _____ of owls.

4. If owls were as big as little children, their eyes would be the size of _____ balls!

6. The _____ in their crop (stomach) is one of the strongest in the natural world, strong enough to kill harmful bacteria in their food.

10. One _____ opening hears sound a split second faster than the other, letting the owl know the location of its prey, how fast it is traveling, and where it is going.

Fill in the blanks from your reading.

1. A group of woodcocks is called a _____ of woodcocks.

2. It takes _____ days for a mother woodcock to teach her babies to fly.

3. During mating season, the male woodcock uses special _____ to serenade the female he chooses.

4. The eyes of woodcocks are in the _____ of their heads.

5. Woodcocks can shove their bills deep into the soil and open them only at the _____.

6. A group of seagulls is called a _____ of seagulls.

7. The _____ of seagulls can be a third of the weight of their entire body.

8. To keep predators away, seagulls make sure to get rid of all the egg shells around their _____ after their babies hatch.

9. Seagulls have a pair of special glands that flush out the _____ they drink.

10. When they live in hot places, seagulls _____ their bodies by keeping their mouths open, like dogs.

Unscramble the missing word or number to fill in the blank.

1. A goose's cone-shaped bill is short and has a single _____ at its end that it uses to pull up grass and chew it. **(hotot)**

2. A group of geese on land is called a _____; in flight, it becomes a skein, team, or wedge of geese. **(gelgag)**

3. If a goose goes missing, its mate goes looking for it, even waiting to _____ until the mate is found. **(eatrgmi)**

4. A group of geese flies _____ percent faster than a goose flying alone. **(07)**

5. The greatest threat to goose eggs are raccoons and _____. **(xosef)**

6. A group of swans in flight is called a _____. **(gewde)**

7. Swans have a protruding _____ in their wings, which is used like an elbow as a weapon. **(nobe)**

8. Swans _____ mainly plants and vegetation that are submerged in the water. **(eta)**

9. Swans have to be lightweight in order to fly, which is why their bones and feather vanes are _____. **(lohowl)**

10. Swans have an oil gland at the base of their tail, which produces oil to keep their _____ water-resistant. **(athresf)**

Fill in the blanks from your reading.

1. Grebes are most famous for their courting dance, known as the "_____."

2. Grebes have to eat _____ every day, and that's why they spend lot of time getting rocks out of the water.

3. Adult grebes molt twice a year, and each time the plumage is a different _____.

4. Unlike other birds, grebes' _____ are denser so that grebes are also much heavier than other birds.

5. A group of grebes is called a water _____ of grebes.

6. Pelican wingspans reach almost _____ feet.

7. Mothers feed their babies by filling their _____ with chewed-up food.

8. Pelicans have _____ pockets under their skin that are used like we use life vests, letting them float and protecting them when they dive in and out of the water.

9. A group of pelicans is called a _____.

10. Colonies flying against the wind fly as _____ as possible.

Crossword Puzzle (Herons and Penguins)

Across

2. At first sight, people thought penguins were a fuzzy type of

4. To dive more easily, penguin bones are not this

5. Herons building nests at different heights all in the same tree create this

8. Heron are found all over the world except here

10. Male penguins make something like this to feed their young

Down

1. If penguin eggs touch the frozen ground, the chick inside die of this

3. A group of heron is called this

6. Heron swallow this part of the fish first because of the bones

7. To take off, herons do this while flapping their wings

9. A group of penguin is called a

Fill in the blanks from your reading.

1. Turkeys can change the color of the skin on their heads or wattles from dark blue to light blue, white, red, or bright red, depending on their _____.

2. In English, they are called "turkeys" because of the _____ merchant who brought them to America for trading.

3. A group of turkeys is called a _____ of turkeys.

4. Jakes have pointed feathers, while adult turkeys have _____ feathers that are often a bit more tattered and worn out.

5. Turkey beards are actually made of bristles that are a type of _____.

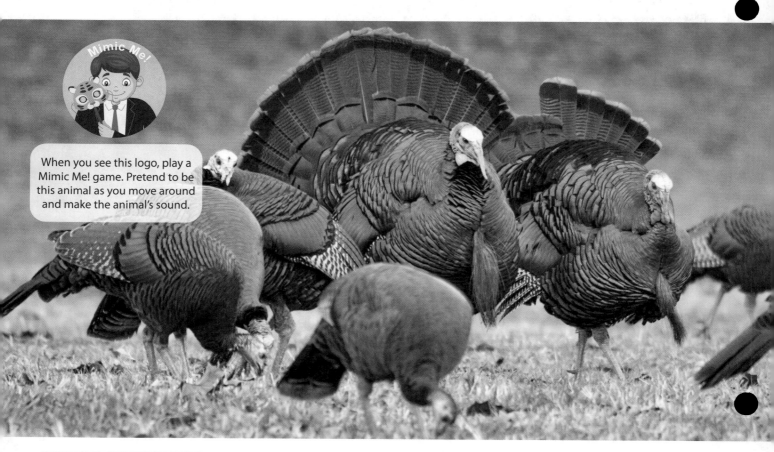

Mimic Me!

When you see this logo, play a Mimic Me! game. Pretend to be this animal as you move around and make the animal's sound.

Unscramble the missing word to fill in the blank.

1. The bright colors of the monarch are used to warn predators that they are _____. **(ouspoonis)**

2. The phase between each monarch molting is called _____. **(arinst)**

3. Monarch butterflies have an annual migration across North America for over _____ miles. **(,5482)**

4. God created monarch butterflies to know which direction they are flying, based on where the _____ is. **(nsu)**

5. A group of butterflies can be called a kaleidoscope of butterflies, a _____ of butterflies, or a rabble of butterflies. **(mwsar)**

6. Unlike many other animals, there is not a specific, official term for a _____ of moths. **(goupr)**

7. Caterpillars have two glands on their heads that secrete a fluid that hardens when it touches _____, forming long threads of silk. **(rai)**

8. The first thing moths do once they break out of the cocoon is _____. **(erst)**

9. Moth _____ that are laid at the end of fall may not survive the winter ahead. **(sgeg)**

10. Caterpillars _____ at least four times because they keep growing. **(lomt)**

Fill in the blanks from your reading.

1. Bees' bodies have three segments: the head, thorax, and _____.

2. Bees eat nothing but the nectar and _____ they collect from flowers.

3. A group of bees is ordinarily called a _____ of bees.

4. Bees let off certain _____ to communicate, which is how the queen gives instructions to the other bees.

5. In summer, bees carry drops of _____ into their hive and rapidly flap their wings to cool themselves.

6. A group of wasps is called a _____ of wasps.

7. God spoke of hornets to describe how He would go before the people of _____ and defeat their enemies.

8. _____ wasps are usually not as aggressive as other wasps and are much less likely to sting people.

9. Wasps first _____ their prey and then shoot venom into the wound they made.

10. God gave the sand wasp the wisdom to gather and get rid of the leftovers by pushing them deeper into the _____.

Crossword Puzzle (Mosquitoes and Flies)

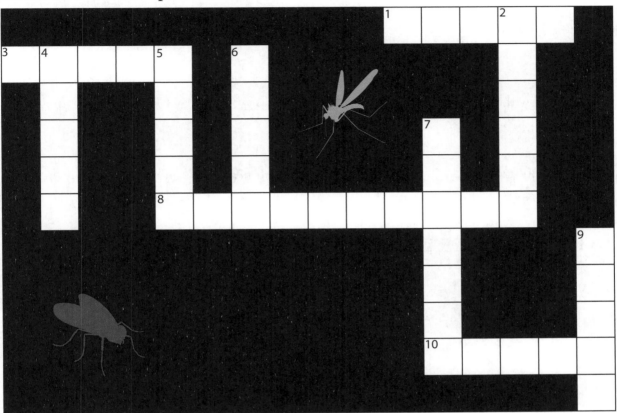

Across

1. Dead flies putrefy the perfumer's ointment, And cause it to give off a foul odor; So does a little _____ to one respected for wisdom and honor. (Ecclesiastes 10:1)

3. House _____ stop all their activities when the temperature drops below 60 degrees or rises higher than 95 degrees.

8. A group of _____ is called a swarm.

10. Mosquitoes are active mainly at _____.

Down

2. Flies have two eyes, but God made each eye to have two thousand _____!

4. A mosquito life cycle consists of four stages: egg, _____, pupa, and adult.

5. A group of flies is called a _____ of flies.

6. Mosquitoes can beat their _____ together up to 600 times in a single second.

7. Female mosquitoes suck on human and animal blood because they need _____ for their unhatched eggs.

9. Flies live an average of a _____.

Fill in the blanks from your reading.

1. A group of fleas is called a _____ of fleas.

2. Fleas are covered with hard plates called _____.

3. Flea larvae feed on traces of _____ that adult fleas leave behind.

4. After making a _____, fleas stick a needle-like organ into the skin.

5. Fleas can jump 100 times higher than their height and a distance of _____ times their length.

6. A group of termites is called a _____ of termites.

7. Termites use their _____ to glue grains of sand together into a hard, sticky mixture and use it to build their nest.

8. All termites serve the _____ and make sure she can do her job of laying eggs.

9. Some termites cultivate the ground in such a way that it causes _____ to grow.

10. God gave termites the wisdom to build their nests with the long sides facing east and _____ so the morning and evening sun can warm their homes.

Fill in the blanks from your reading.

1. A group of frogs is called an _____ of frogs.

2. The deadliest type of poison dart frog is called the _____ poison dart frog.

3. Most frogs are nocturnal, but these frogs are diurnal, which means they are most active during the

   _____.

4. The poison dart frogs are considered poisonous because they do not have to _____ to

   be dangerous.

5. As a general rule, younger frogs are much _____ poisonous than older ones.

6. A group of turtles can be called by several names, including a bale of turtles, a dule of turtles, a nest of

   turtles, and a _____ of turtles.

7. The sun helps turtles warm up and rid themselves of _____.

8. Common box turtles live for an average of 40 years, but some species live to be _____ or more.

9. Turtles have strong, jagged _____ that help them take bites.

10. When the temperature gets cold, turtles simply go to _____.

Unscramble the missing word to fill in the blank.

1. A group of alligators is called a _____ of alligators.
   (grecongraiont)

2. Alligators live in _____ and in the United States. **(aCihn)**

3. An alligator can eat a whole _____ and are powerful enough to kill one quickly.
   (ocw)

4. The longer an alligator spends time in water with lots of _____, the greener it becomes.
   (eagla)

5. The lower eyelids _____ the alligators' eyes when underwater. **(otprcte)**

6. The Komodo dragon can be up to _____ feet long. **(01)**

7. The people of Indonesia call Komodo dragons by the name *ora*, which means land
   _____. **(lecrodico)**

8. A group of Komodo dragons are sometimes called a _____ of Komodo dragons.
   (nkab)

9. Komodo dragons use their _____ to smell, which is why they are always flicking their tongues out. **(uetong)**

10. The older Komodo dragons will eat the babies, so the babies crawl into _____ to protect themselves. **(ertes)**

Fill in the blanks from your reading.

1. Because marine iguanas are _____-blooded, like other reptiles, their body temperatures depend on the outside temperature.

2. Marine iguanas are the only iguana that likes spending time in and near the _____.

3. Marine iguanas are currently classified as _____.

4. Marine iguanas have been known to shrink as much as _____% in their length when food is scarce.

5. A group of iguanas is called a _____ of iguanas.

6. Chameleon feet are designed to work like _____, which also helps them hold their place on a tree branch.

7. There is not a special term for a group of chameleons, but a group of lizards is called a _____ of lizards.

8. Chameleons have special _____ in a layer under their skin that reflect and absorb light, and these move closer together or farther apart when they change color.

9. Chameleons have to _____ their tongues in their mouths accordion-style.

10. Chameleon color changes occur due to the lizard's _____ or the environment.

Fill in the blanks from your reading.

1. When snakes want to swallow a whole egg, they disjoint their _____ to have

    maximum space in their mouths.

2. A group of rattlesnakes is called a _____ of rattlesnakes.

3. Snakes have an inner-_____ they use to feel vibrations in the ground.

4. Rattlesnakes are vipers that use their _____ on rabbits, rats, and birds.

5. Venom is a strong _____ that some snakes inject into their prey, and it can cause death.

Fill in the blanks from your reading.

1. Musk deer do not have antlers, but instead grow long _____.

2. A group of deer is called a _____ of deer.

3. Deer hooves are as sharp as _____.

4. Deer have glands on their feet that secrete a kind of scented _____.

5. At the end of the breeding season, deer shed their _____.

6. The single-humped camel, known as the _____, is faster than the two-humped camel.

7. Most Bactrian camels live in _____ and Mongolia and are used as pack animals.

8. Camels have special skin on their knees and _____ that protects them from the heat.

9. Camels use their eyelashes as a strainer, keeping _____ from getting into their eyes.

10. A group of camels is called a _____ of camels.

Fill in the blanks from your reading.

1. Elephants are the largest _____ on land.

2. To survive, elephants require over _____ pounds of food and nearly 40 gallons of water every day.

3. Baby elephants drink _____ gallons of milk every day.

4. A group of elephants is called a _____ of elephants.

5. Elephants use their trunks to cover themselves with _____, mainly to protect themselves from insect bites.

6. A full-grown gorilla stands well over 6 feet tall and weighs about _____ pounds.

7. A group of gorillas is called a _____ of gorillas.

8. Playing and _____ are some of the ways that gorilla family group members bond with each other.

9. Evolutionists falsely claim that humans and gorillas come from a "common _____."

10. Gorillas live in family groups that can have anywhere from 10–_____ gorillas.

Unscramble the missing word to fill in the blank.

1. In addition to their regular diet of plants, rabbits also eat their own _____ to nourish themselves and get protein and vitamins they need to be healthy. **(ppdrogsin)**

2. Rabbits have scent glands on their _____ and rub them on things to mark their territory. **(iscnh)**

3. Groups of rabbits can be called several names, including a colony of rabbits, a bury of rabbits, and a _____ of rabbits. **(tnse)**

4. Rabbits' breeding season starts in January and ends in _____. **(uJen)**

5. Rabbits have a very keen sense of _____, which they need for survival. **(emlsl)**

6. Opossums play _____ to defend themselves from humans and other predators. **(ddae)**

7. A male opossum is a jack, a female is a jill, and a baby is a _____. **(yoje)**

8. Opossums are born very tiny, about half the size of a _____. **(ebe)**

9. Tiny opossums lose their _____. **(slcaw)**

10. Opossums leave their mother's _____ after 60 or 70 days. **(uchop)**

Fill in the blanks from your reading.

1. Shrews are the smallest _____ in the world.

2. The word "shrew" comes from the Old English word _____, which means old man or dwarf.

3. A shrew's main food source is insects such as beetles, ants, flies, and _____.

4. Dozens of shrews can live separately on an acre of land, each one with his or her own

    _____.

5. Shrews have a _____ gland in their mouth that can kill 200 mice in one bite.

6. Usually, they are _____, but when necessary, they also eat snails or other

    dead mice.

7. A group of mice is called a _____ of mice.

8. We can estimate the age of mice based on their _____.

9. Mice gather all kinds of materials to make their homes cozy, including paper, cardboard boxes, mattress

    stuffing, and even _____.

10. Mice sleep in a group in winter, with the weakest, smallest mouse in the _____ circle.

Fill in the blanks from your reading.

1. Some squirrels are very friendly and live in a group called a _____, but most are solitary.

2. A group of squirrels is called a _____ of squirrels.

3. God specially designed squirrel _____ to work perfectly for their life in the trees.

4. A squirrel eats more than _____ pounds of seeds, fruit, acorns, mushrooms, and insects per year.

5. Squirrels prefer rubbing their faces against _____ instead of using their paws to clean up.

Help the groundhog get to his burrow.

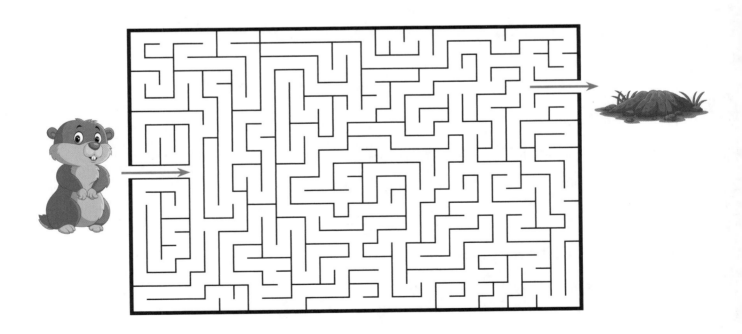

Fill in the blanks from your reading.

1. Beaver fossils in England show that, in the past, beavers could grow to over _____ feet.

2. Beavers have a pair of big, strong _____ that grow all the time.

3. In times of danger, beavers slap the water with their _____, making a loud noise.

4. When they build _____, deep water pockets are created where beavers can dive for protection when being attacked.

5. A group of beavers is called a colony of beavers or a _____ of beavers.

6. The word porcupine comes from French and means "thorny _____."

7. Porcupines have _____ that control each one of its quills.

8. Porcupines have quills, but God made sure the quills are _____ so they don't hurt the mother during birth.

9. When porcupines feel under attack, they turn their _____ toward their enemies and puff out their quills to look bigger and scarier.

10. Porcupines do not live in big groups, but keep to their own _____.

Fill in the blanks from your reading.

1. The strong scent skunks _____ can keep even a big bear from attacking.

2. A group of skunks is called a _____ of skunks.

3. Skunks will provide _____ before they spray by stamping their feet or growling.

4. The only animals unaffected by skunk spray are _____.

5. Skunks are mainly active at night and are very helpful to _____ because they eat mice, rats, squirrels, locusts, beetles, and other insects that damage crops.

6. The word "raccoon" came from a native _____ term that means "the one who rubs, scrubs, and scratches with its hands."

7. Raccoons are one of the most common animals to carry _____, though they also can carry other parasites and diseases.

8. Raccoons are _____, so you can see them at night when they are out searching for food.

9. Male raccoons are called boars, female raccoons are called sows, and baby raccoons are called _____.

10. Raccoons _____ their food in water before eating it.

Crossword Puzzle (Badgers and Otters)

Across

2. Badgers _____ while looking for food such as insects, seeds, roots, vegetables, fruits, and nuts.

3. When otters dive or swim, their _____ close, keeping water out of their noses and ears.

7. Badgers dig little pits and use them as _____.

8. When otters swim in murky waters, they use their whiskers to feel how _____ the water is, how far they are from shore, or if there are any fish moving nearby.

9. Otters can see underwater just as well as on land, even though light rays are _____ by the water.

Down

1. Badgers are thought to be the bravest creatures in Africa — even _____ stay away from them.

4. A group of otters can be called a _____ of otters or a raft of otters.

5. A group of badgers is called a _____ of badgers.

6. Most animals who live above the ground have their fur facing one way, but badgers have fur that can _____ in both directions.

8. When an otter _____, the rest of the family gathers around him and stays put for days without eating or playing.

Fill in the blanks from your reading.

1. Because weasels can be so _____, some animals freeze or even die of fear when coming across a weasel.

2. A group of weasels can be called a gang of weasels, a pack of weasels, a sneak of weasels, a confusion of weasels, and even a _____ of weasels.

3. Weasels' strength is in their _____ and courage.

4. Weasels have two coats of _____, one for winter and one for summer.

5. Weasels are known for the war _____ they do when they see prey.

6. Meerkats have different shades of color on their _____, which they use to recognize one another.

7. Meerkat comes from Dutch and means "lake _____."

8. Gang life for meerkats is very organized, which means there are _____ that must never be broken or those who break them are punished.

9. A group of meerkats is called a _____ of meerkats or a gang of meerkats.

10. The most meerkats a gang will have is about _____.

Unscramble the missing word to fill in the blank.

1. Foxes are one of the most common carriers of _____. **(iebsra)**

2. A group of foxes is called a skulk of foxes or a _____ of foxes. **(shlea)**

3. Foxes wrap their tails around themselves like blankets and cover their feet and snout because those are the areas that are not covered in _____. **(rfu)**

4. God gave foxes the ability to keep their rhinariums moist because the moisture helps make their sense of _____ stronger. **(lemsl)**

5. Foxes take turns chasing after their _____ so that one can rest while the other continues the chase. **(reyp)**

6. Wolves have _____ toes on their front paws and four toes on their back paws. **(vfei)**

7. God created wolves with small bodies and strong _____, which is why they can run fast and far. **(sgle)**

8. The _____ of a wolf are so strong that they can crush bones. **(aswj)**

9. When wolves get hungry, they hunt _____. **(theogetr)**

10. A group of wolves is called a _____ of wolves. **(pcak)**

Fill in the blanks from your reading.

1. Lions can run up to _____ miles per hour over short distances and jump as high as 12 feet.

2. A group of lions is called a _____ of lions.

3. African lions mainly live in the _____, which are called savannahs.

4. Lions are the most _____ of cats.

5. Young male lions often have to leave their family pride once they start growing a _____.

6. Tigers are the largest _____ in the world, and the third-largest carnivores on land.

7. Bengal tigers are found in _____ and are the most common type of tiger, though they are endangered.

8. Tigers are generally _____ animals.

9. A group of tigers is called a streak of tigers or an _____ of tigers.

10. Tigers will sometimes save their food for later by _____ it.

Fill in the blanks from your reading.

1. Adult Alaskan grizzly bears can, however, weigh more than _____ pounds.

2. Grizzly bears are the largest _____ on land.

3. A group of bears in general is called a sloth of bears or a _____ of bears.

4. Cubs stay close to their mom until the age of _____.

5. During a bear's _____, their bodies rest and heal from all the injuries they got while looking for food during the summer months.

6. Bats are flying _____.

7. The _____ bat is the only bat that does not feed on insects.

8. Bats can eat up to a _____ of their body weight in insects in a single night!

9. A group of bats is called a _____ of bats.

10. Bats use a system called echolocation, which locates objects using _____.

Fill in the blanks from your reading.

1. Dolphins can dive as deep as 305 feet, jump over 6 feet high, and swim faster than _____ miles per hour.

2. Dolphins have up to _____ teeth, but they like to swallow their food whole without chewing it.

3. The word "dolphin" comes from the word _____ in Greek, which means "fish with a womb."

4. Dolphins have two lobes called flukes on their tails, which they use for swimming and _____.

5. A group of dolphins is called a _____.

Fill in the blanks from your reading.

1. Beluga whales are nicknamed "white whales" because of their color and "sea canaries" because of their

   _____.

2. The name "beluga" comes from the Russian word for "white": _____.

3. A group of whales can be called several names, including a pod of whales, a school of whales, or a

   _____ of whales.

4. Beluga whales are huge animals, but nearly half of their weight comes from their

   _____ (fat).

5. The calves are usually born in the late spring or during the summer when the water and the

   temperatures overall are _____.

Fill in the blanks from your reading.

1. Salmon hatch in fresh water (usually in rivers) and migrate to _____ water.

2. In Alaska, the arrival of salmon indicates _____.

3. Salmon swim long distances to reach the place where they _____.

4. After spawning during the _____, eggs stay in the ground on the river bank.

5. When the little salmon _____, they swim for many miles to reach the ocean and will one day return to the same spot they came from.

Fill in the blanks from your reading.

1. Great white sharks are the largest _____ fish in the ocean.

2. Great white sharks generally stay away from each other because they will _____ each other!

3. A great white shark's electroreception abilities are so sensitive that they can detect another animal's _____ from a few hundred feet away.

4. A group of sharks is called a school of sharks or a _____ of sharks.

5. Sharks are found off the coasts of all continents except _____.

Fill in the blanks from your reading.

1. The mouth of an octopus has a sharp, hard beak, the only hard _____ in their bodies.

2. Little baby octopuses start hatching after eight to ten weeks, depending on the water's

    _____.

3. Octopuses have three layers of color cells, called chromatophores, just below the surface of their skin:

    red, blue, and _____.

4. Octopuses can "_____" arms when under attack.

5. Babies are called larvae, hatchlings, or _____.

Fill in the blanks from your reading.

1. A jellyfish is an invertebrate, which means they do not have a _____.

2. The cells that release venom when a jellyfish is eating are what cause jellyfish _____.

3. There are several names for a group of jellyfish, including a bloom of jellyfish, a smack of jellyfish, or a _____ of jellyfish.

4. Symmetrization means that jellyfish can grow back features and _____ and even reorganize their new legs to make sure that both sides were like they were before an injury or accident.

5. Jellyfish do not have many organs that we think of as common, including a brain or a _____.

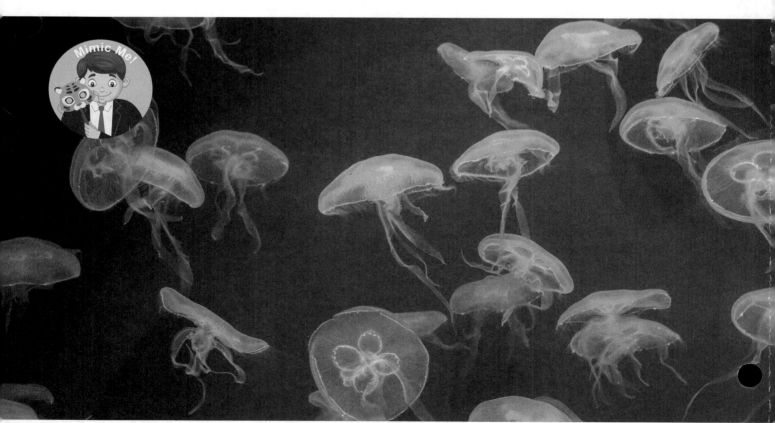

Mimic Me!

Worksheets

for Use with

The Complete Zoo Adventure

Color the picture of the toucan below.

Color the picture of the lion's head below.

Draw a line to match the correct eyes God has created for each animal.

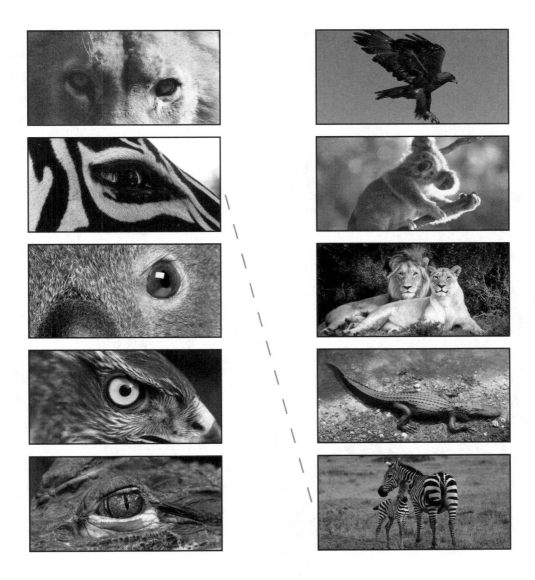

Connect the dots.

CREATED
(practice upper case)

created
(practice lower case)

created
(practice cursive)

OWL
(practice upper case)

owl
(practice lower case)

owl
(practice cursive)

Find and color the birds in the picture below.

Find these words in the puzzle below.

1. Chimpanzee
2. Horses
3. Komodo Dragon
4. Frogs
5. Fall
6. Chaparral
7. Purpose
8. Taiga
9. Flamingo
10. Wolf
11. Alligator
12. Biome
13. Creator
14. Rainforests
15. Migration
16. Elephant
17. Macaw
18. Soil
19. Christ
20. Journal
21. Tundra
22. Polar Bear
23. Deserts
24. Camel
25. Zoo
26. Ark
27. Adventure
28. Hummingbird
29. Water
30. Kangaroo

T	U	N	D	R	A	F	C	C	A	D	K	A	N	G	A	R	O	G	H
U	R	A	B	V	T	L	H	P	X	T	U	Z	M	S	Z	S	M	W	J
N	A	B	D	N	A	A	R	N	O	G	A	R	D	O	D	O	M	O	K
D	I	D	Q	V	I	M	I	L	Z	K	I	I	B	D	X	I	Q	L	A
G	N	O	I	B	E	I	S	K	A	L	M	I	G	R	A	L	W	F	N
L	F	X	W	I	B	N	T	M	S	Q	O	Z	V	A	C	O	E	H	G
H	O	Z	E	O	G	N	I	M	A	L	F	X	V	D	V	P	R	U	A
O	R	Q	A	M	A	C	A	H	A	P	A	R	R	A	L	T	M	R
F	E	M	N	D	N	W	R	N	L	W	B	E	L	R	L	I	Y	M	O
R	S	J	O	U	R	N	A	L	D	O	O	Z	I	L	L	O	U	I	O
O	T	E	N	K	M	E	T	W	A	C	A	M	N	G	I	D	I	N	C
S	S	C	S	L	Q	W	A	T	E	R	P	C	J	H	G	R	Z	G	H
T	Z	A	F	R	O	G	S	K	F	E	E	V	B	J	A	T	O	B	R
R	T	M	G	Z	O	C	Y	B	R	R	W	S	C	I	T	Y	P	I	S
E	Y	E	H	X	C	H	I	M	P	A	N	Z	E	E	O	U	A	R	T
S	U	L	J	C	V	A	I	V	G	T	E	B	X	D	R	M	S	D	S
E	S	O	P	R	U	P	O	C	H	T	N	A	H	P	E	L	E	K	Q
D	I	F	P	O	L	A	R	B	E	A	R	N	Z	K	B	I	D	R	A
P	O	A	S	D	C	R	E	A	T	O	R	M	A	L	N	O	F	L	Z
E	R	U	T	N	E	V	D	A	J	Y	N	O	I	T	A	R	G	I	M

Draw a line to match the correct ears God has created for each animal.

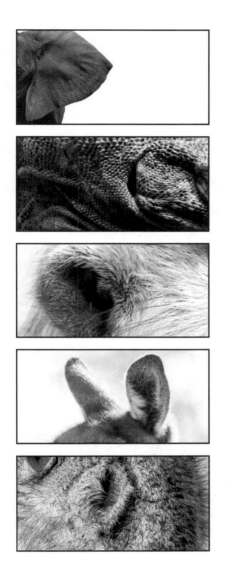

Color the cockatoo.

Start at number 1 and connect the dots. Color the picture.

Using your DESERT biome card in the folder in the back of your book, write about or draw pictures of three things you learned about deserts!

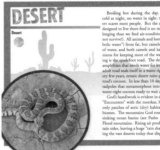

1.

2.

3.

Draw a line to match the correct nose God has created for each animal.

Start at number 1 and connect the dots.

PAWS
(practice upper case)

paws
(practice lower case)

paws
(practice cursive)

TUNDRA
(practice upper case)

tundra
(practice lower case)

tundra
(practice cursive)

Find and color these mammals with paws in the picture below.

Find these words in the puzzle below.

1. Peacock
2. Creation
3. Hawk
4. Grasslands
5. Giraffe
6. Climate
7. Ararat
8. Zebra
9. Owl
10. Hippopotamus
11. Koala
12. Kind
13. Panda
14. Parrot
15. Mountainous
16. Crocodile
17. Habitat
18. Flood
19. Fruit Bat
20. Hyrax
21. Plan
22. Meerkat
23. Lion
24. Lowland
25. Garden
26. Eagle
27. Rhinoceros
28. Design
29. Gorilla

B	A	T	Q	W	E	R	C	L	I	M	A	T	E	T	E	L	G	A	E
A	H	Z	X	C	B	K	C	O	C	A	E	P	N	G	I	S	E	D	G
S	A	L	O	W	A	L	L	I	R	O	G	L	D	F	Y	U	I	O	R
A	B	T	N	A	L	S	S	A	R	G	M	A	K	Q	L	M	L	P	A
W	I	Y	V	Q	X	L	A	N	D	G	B	N	W	E	I	O	K	H	S
O	T	D	E	S	I	A	R	A	R	A	T	T	R	T	Y	U	O	J	S
P	A	N	D	A	L	W	O	K	P	A	R	E	O	I	O	N	A	D	L
M	T	U	C	W	K	O	A	L	A	S	V	F	L	R	P	T	B	R	A
R	E	I	X	G	I	R	A	F	R	D	C	X	G	O	R	A	C	I	N
H	P	E	A	G	N	A	K	W	A	H	C	I	B	A	H	I	P	N	D
I	L	O	R	E	N	M	D	N	I	K	R	Z	O	V	F	N	G	O	S
N	G	I	R	A	F	F	E	Z	E	B	O	O	L	F	L	O	A	C	J
O	O	A	O	N	O	I	T	A	E	R	C	C	N	R	O	U	R	E	H
C	R	D	Z	N	T	A	E	R	C	L	O	M	A	U	D	S	D	R	P
E	R	F	E	R	L	O	W	L	A	N	D	B	M	I	C	G	E	O	A
R	I	G	B	T	H	Y	P	A	X	S	I	V	F	T	L	F	N	S	R
O	H	G	R	D	A	N	L	H	G	F	L	I	O	B	I	D	S	A	R
S	E	B	A	C	R	E	A	T	I	D	E	G	X	A	R	Y	H	Z	O
A	Q	G	N	A	P	T	A	K	R	E	E	M	B	T	M	X	C	V	T
Z	X	C	H	I	P	P	O	P	O	T	A	M	U	S	A	B	N	M	C

Draw a line to match the correct mouth God has created for each animal.

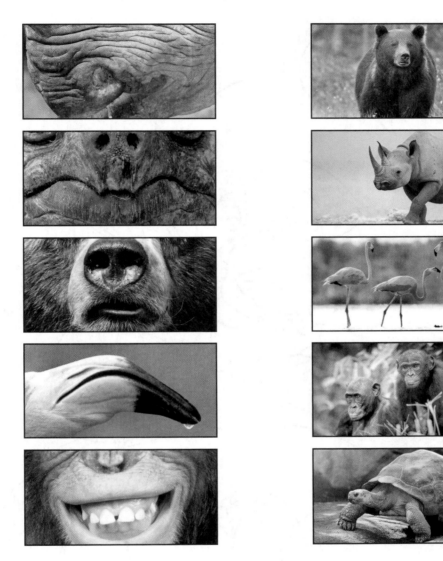

Color the panda below.

Using your GRASSLAND biome card in the folder in the back of your book, write about or draw pictures of three things you learned about grasslands!

1.

2.

3.

Using your TUNDRA biome card in the folder in the back of your book, write about or draw pictures of three things you learned about the tundra!

1.

2.

3.

Draw a line to match the correct foot God has created for each animal.

Using your CHAPARRAL biome card in the folder in the back of your book, write about or draw pictures of three things you learned about chaparrals!

1.

2.

3.

Color the meerkats below.

Using your RAINFOREST, EVERGREEN FOREST, and DECIDUOUS FOREST biome cards in the folder in the back of your book, write about or draw pictures of three things you learned about forests!

1.

2.

3.

Cut out and match the head and tail.

For each animal, describe your favorite design feature (adaptations):

1. Flamingo	
2. Peacock	
3. Hummingbird	
4. Parrot & Macaw	
5. Eagles & Hawks	
6. Owl	
7. Fruit Bat	
8. Arctic Wolf	
9. Polar Bear	
10. Panda	
11. Koala	
12. Kangaroo	
13. Lion	
14. Meerkat	
15. Hyrax	
16. Chimpanzee	
17. Gorilla	
18. Giraffe	
19. Zebras	
20. Camel	
21. Elephant	
22. Rhinoceros	
23. Hippopotamus	
24. Komodo Dragon	
25. Alligators & Crocodiles	
26. Tortoise	
27. Tree Frog	

God has provided great variety in the creation!

Give the day each animal was created and look up and write the Scriptures for Day 5 and 6. See your animal cards in the back of your zoo book.

1. Flamingo	
2. Peacock	
3. Hummingbird	
4. Parrot & Macaw	
5. Eagles & Hawks	
6. Owl	
7. Fruit Bat	
8. Arctic Wolf	
9. Polar Bear	
10. Panda	
11. Koala	
12. Kangaroo	
13. Lion	
14. Meerkat	
15. Hyrax	
16. Chimpanzee	
17. Gorilla	
18. Giraffe	
19. Zebras	
20. Camel	
21. Elephant	
22. Rhinoceros	
23. Hippopotamus	
24. Komodo Dragon	
25. Alligators & Crocodiles	
26. Tortoise	
27. Tree Frog	

Day 5 – Genesis 1:20–23:

Day 6 – (in part) Genesis 1:24–27, 31:

God created ALL THINGS in SIX DAYS!

Start at number 1 and connect the dots.

HOOVES
(practice upper case)

hooves
(practice lower case)

hooves
(practice cursive)

GIRAFFE
(practice upper case)

giraffe
(practice lower case)

giraffe
(practice cursive)

Cut out on the lines and then match.

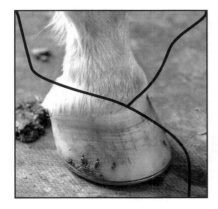

Hooves (Horse)

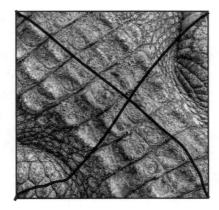

Scutes (Alligator)

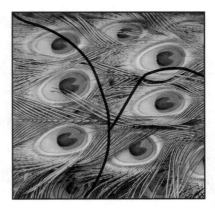

Feathers (Peacock)

Fur (Meerkat)

Beak (Macaw)

Claws (Koala)

Find and color the mammals with hooves in the picture below.

Give the genus and species of these 27 animals.

Animal	Genus	Species
1. Flamingo	1.	1.
2. Peacock	2.	2.
3. Hummingbird	3.	3.
4. Parrot & Macaw	4.	4.
5. Eagles & Hawks	5.	5.
6. Owl	6.	6.
7. Fruit Bat	7.	7.
8. Arctic Wolf	8.	8.
9. Polar Bear	9.	9.
10. Panda	10.	10.
11. Koala	11.	11.
12. Kangaroo	12.	12.
13. Lion	13.	13.
14. Meerkat	14.	14.
15. Hyrax	15.	15.
16. Chimpanzee	16.	16.
17. Gorilla	17.	17.
18. Giraffe	18.	18.
19. Zebras	19.	19.
20. Camel	20.	20.
21. Elephant	21.	21.
22. Rhinoceros	22.	22.
23. Hippopotamus	23.	23.
24. Komodo Dragon	24.	24.
25. Alligators & Crocodiles	25.	25.
26. Tortoise	26.	26.
27. Tree Frog	27.	27.

Label the following animals as Nocturnal, Arboreal, Browsers, or Grazers. See your animal cards in the back folder of your zoo book.

1. Flamingo	
2. Peacock	
3. Hummingbird	
4. Parrot & Macaw	
5. Eagles & Hawks	
6. Owl	
7. Fruit Bat	
8. Arctic Wolf	
9. Polar Bear	
10. Panda	
11. Koala	
12. Kangaroo	
13. Lion	
14. Meerkat	
15. Hyrax	
16. Chimpanzee	
17. Gorilla	
18. Giraffe	
19. Zebras	
20. Camel	
21. Elephant	
22. Rhinoceros	
23. Hippopotamus	
24. Komodo Dragon	
25. Alligators & Crocodiles	
26. Tortoise	
27. Tree Frog	

Give each animal's group: Reptile, Mammal, Bird, and Amphibian. (See "Looking Ahead — 5," page 21, *The Complete Zoo Adventure*.) Then define reptile, mammal, bird, and amphibian in your own words. See your animal cards in the back folder of your zoo book.

1. Flamingo	
2. Peacock	
3. Hummingbird	
4. Parrot & Macaw	
5. Eagles & Hawks	
6. Owl	
7. Fruit Bat	
8. Arctic Wolf	
9. Polar Bear	
10. Panda	
11. Koala	
12. Kangaroo	
13. Lion	
14. Meerkat	
15. Hyrax	
16. Chimpanzee	
17. Gorilla	
18. Giraffe	
19. Zebras	
20. Camel	
21. Elephant	
22. Rhinoceros	
23. Hippopotamus	
24. Komodo Dragon	
25. Alligators & Crocodiles	
26. Tortoise	
27. Tree Frog	

Reptile:

Amphibian:

Bird:

Mammal:

God has created DIFFERENCES!

List each animal (from the animal cards in the back folder of your zoo book) under its ecological area.

Grassland:
Hardwood (Deciduous) Forest:
Conifer (Evergreen) Forest (Taiga):
Tropical Rain Forest:
Desert:
Tundra:
Chaparral:

God's creatures fill the earth!

Start at number 1 and connect the dots.

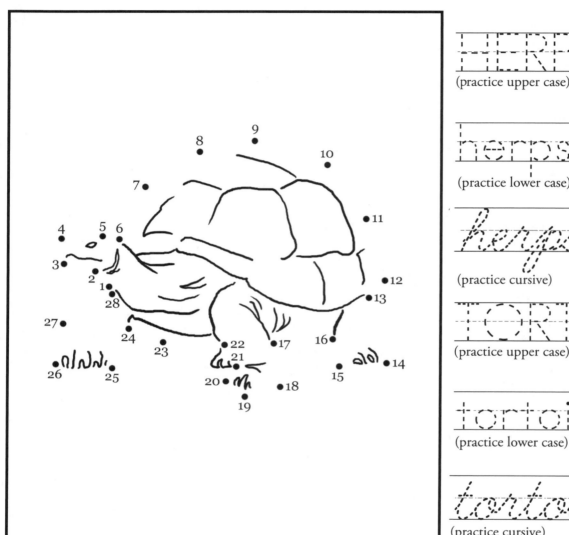

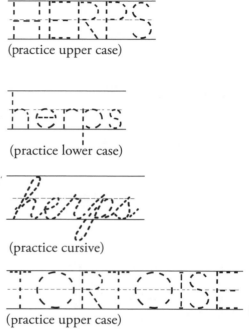

HERPS
(practice upper case)

herps
(practice lower case)

herps
(practice cursive)

TORTOISE
(practice upper case)

tortoise
(practice lower case)

tortoise
(practice cursive)

Find and color the reptiles and amphibians in the picture below.

List each animal's typical diet. See your animal cards in the back folder of your zoo book.

1.	Flamingo	
2.	Peacock	
3.	Hummingbird	
4.	Parrot & Macaw	
5.	Eagles & Hawks	
6.	Owl	
7.	Fruit Bat	
8.	Arctic Wolf	
9.	Polar Bear	
10.	Panda	
11.	Koala	
12.	Kangaroo	
13.	Lion	
14.	Meerkat	
15.	Hyrax	
16.	Chimpanzee	
17.	Gorilla	
18.	Giraffe	
19.	Zebras	
20.	Camel	
21.	Elephant	
22.	Rhinoceros	
23.	Hippopotamus	
24.	Komodo Dragon	
25.	Alligators & Crocodiles	
26.	Tortoise	
27.	Tree Frog	

Start at number 1 and connect the dots.

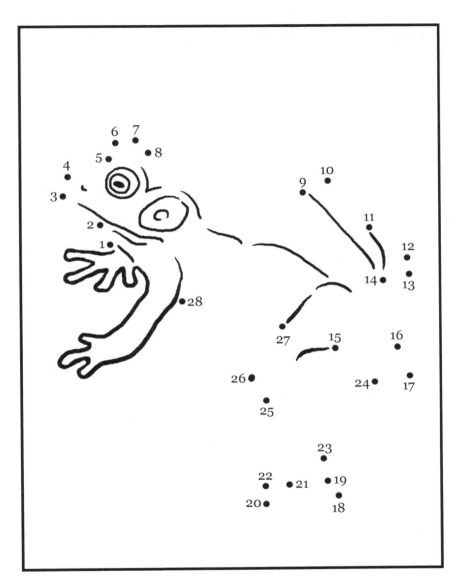

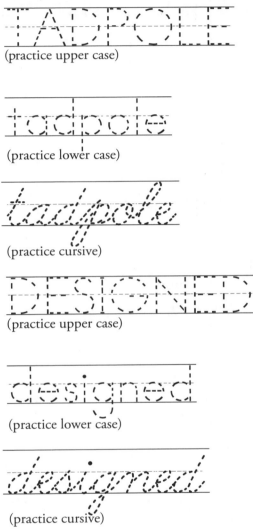

TADPOLE
(practice upper case)

tadpole
(practice lower case)

tadpole
(practice cursive)

DESIGNED
(practice upper case)

designed
(practice lower case)

designed
(practice cursive)

Give the length (sometimes it is actually the height), weight, and life span of these 27 animals. See your animal cards in the back folder of your zoo book.

Animal	Length (body)	Weight (maximum)	Life Span (in wild)
1. Flamingo	1.	1.	1.
2. Peacock	2.	2.	2.
3. Hummingbird	3.	3.	3.
4. Parrot & Macaw	4.	4.	4.
5. Eagles & Hawks	5.	5.	5.
6. Owl	6.	6.	6.
7. Fruit Bat	7.	7.	7.
8. Arctic Wolf	8.	8.	8.
9. Polar Bear	9.	9.	9.
10. Panda	10.	10.	10.
11. Koala	11.	11.	11.
12. Kangaroo	12.	12.	12.
13. Lion	13.	13.	13.
14. Meerkat	14.	14.	14.
15. Hyrax	15.	15.	15.
16. Chimpanzee	16.	16.	16.
17. Gorilla	17.	17.	17.
18. Giraffe	18.	18.	18.
19. Zebras	19.	19.	19.
20. Camel	20.	20.	20.
21. Elephant	21.	21.	21.
22. Rhinoceros	22.	22.	22.
23. Hippopotamus	23.	23.	23.
24. Komodo Dragon	24.	24.	24.
25. Alligators & Crocodiles	25.	25.	25.
26. Tortoise	26.	26.	26.
27. Tree Frog	27.	27.	27.

Quiz Section

for Use with

Elementary Zoology

Aquarium Animal Crossword

ACROSS
2. This white whale lives in the Arctic Ocean.
4. The _____ dolphin is common to many aquariums and always appears to be smiling.
5. The _____ whale is also called "orca."
6. The ocean off the west coast of America.
11. The book of Genesis says that God created everything in _____ days.
13. This bird dives in water and then comes ashore to let its feathers dry out.
14. This sea turtle has a large head like a log.
16. This sea bird has a huge throat pouch.
17. The sea _____ can grow back an arm if it loses it.
18. This sea invertebrate has 8 arms.
20. A type of marine life park where sea creatures live.
21. This shark was not named after someone in the medical field.

DOWN
1. This animal is 95% water.
3. This "toothy" reptile has leathery skin and lives in southern swamps and lakes.
7. The ocean off the east coast of America.
8. The sea _____ is a reptile and lays its eggs up on the sand.
9. This sea creature has a flattened body and spends time on the bottom feeding.
10. According to the Bible, God created sea animals on day _____.
11. This fish is from Australia and has fins that look like seaweed.
12. The male _____ gives birth to the babies.
15. This fish resembles a huge pincushion when disturbed.
19. The bonnethead is an example of this animal.

Animal Observations

Observe and put a check mark for every feature each animal has. Write the number of legs each animal has in the No. of Legs column.

	Hooves	Pouch	Fur/Hair	Scales	Feathers	Claws	No. of Legs	Wings
1. Flamingo								
2. Peacock								
3. Hummingbird								
4. Parrot & Macaw								
5. Eagles & Hawks								
6. Owl								
7. Fruit Bat								
8. Arctic Wolf								
9. Polar Bear								
10. Panda								
11. Koala								
12. Kangaroo								
13. Lion								
14. Meerkat								
15. Hyrax								
16. Chimpanzee								
17. Gorilla								
18. Giraffe								
19. Zebras								
20. Camel								
21. Elephant								
22. Rhinoceros								
23. Hippopotamus								
24. Komodo Dragon								
25. Alligators & Crocodiles								
26. Tortoise								
27. Tree Frog								

Bonus Activity

Green Sea Turtle Hand Puppet

Copy, color in the green sea turtle, then cut out the form. Fold and tape along the dotted lines to finish the puppet.

One way to show concern for sea turtles is to be careful when disposing of plastic bags or balloons. Sea turtles don't see well and often mistake these items for jellyfish — one of their favorite foods! If swallowed, a plastic bag or balloon can hurt or kill a sea turtle!

Bonus Activity

Sea Scramble 1

Copy, cut out, and then arrange in the correct order to make a complete sea star picture.

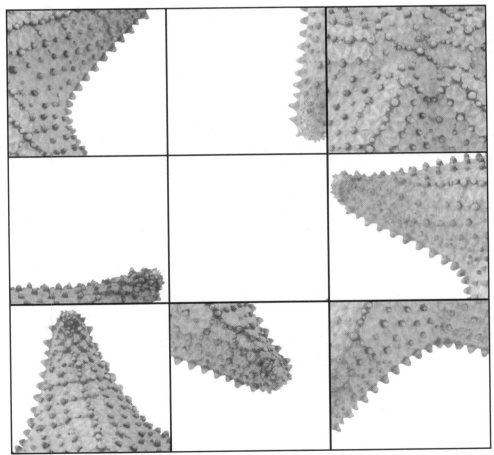

Let all things be done decently and in order.
—1 Corinthians 14:40

Bonus Activity

Sea Scramble 2

Copy, cut out, and then arrange in the correct order to make a complete killer whale picture.

He hath made every thing beautiful in his time: also he hath set the world in their heart,
so that no man can find out the work that God maketh from the beginning to the end.

—Ecclesiastes 3:11 (KJV)

Appendix One

Aquarium Field Trip or Aquarium in the House Day

Aquarium Field Trip: If you will be going to visit an aquarium, *The Complete Aquarium Adventure* is created just for that purpose:

- Start with the Chart Your Course list on page 9.

- Read through the Aquarium Adventure Manifest on page 11.

- Look over the At the Aquarium pages 29–33 to make sure everyone is ready for the trip.

- Make sure you go over the All Ashore page (112) to reflect on your trip when you get back.

- Enjoy your special day!

Aquarium in the House Day: If you will be creating your own aquarium trip without going anywhere, prepare your home or class for adventure:

- You can certainly go over the same introductory pages as above. Gather all the books you have about animals to look through over the day. You might even set any stuffed animals you have around the house for fun!

- Consider making special snacks and a special lunch for this special day of activities. Keep hydrated!

- In the back pocket of *The Complete Aquarium Adventure* are the fish, bird, and animal cards. A teacher can hide these in various rooms in envelopes so you can't see what they are until opened. Then use the BingOcean cards in the pocket to do a search and find, seeking five in a row for a win. See how fast you can get a BingOcean!

- Once your search and find is done (and you can certainly do it more than once) gather up all the cards to see who can answer the questions about the creatures.

- Make the day a special one!

Appendix Two

Zoo Field Trip or Zoo in the House Day

Zoo Field Trip: If you will be going to visit a zoo, *The Complete Zoo Adventure* is created just for that purpose:

- Start by looking over your Planning Calendar list on pages 10–11.

- Read through the Field Trip Prep Time on page 26.

- Look over the At the Zoo pages 27–29 to make sure everyone is ready for the trip.

- Make sure you go over the Around the Campfire page (104) to reflect on your trip when you get back.

- Enjoy your special day!

Zoo in the House Day: If you will be creating your own zoo trip without going anywhere, prepare your home or class for adventure:

- You can certainly go over the same introductory pages as above. Gather all the books you have about animals to look through over the day. You might even set any stuffed animals you have around the house for fun!

- Consider making special snacks and a special lunch for this special day of activities. Keep hydrated!

- In the back pocket of *The Complete Zoo Adventure* are the animal and bird cards. A teacher can hide these in various rooms in envelopes so you can't see what they are until opened. Then the teacher can call out a creature for you to find and all students need to race to see who can find that one first. Fill up the Field Journal cards that are in the back pocket as well!

- Once your search and find is done, gather up all the cards to see who can answer the most questions about the creatures.

- Make the day a special one!

Answer Keys

for Use with

Elementary Zoology

How Many Animals Were on the Ark? → Worksheet Answer Keys

Worksheet 2

1. Ten
2. living
3. features
4. kind
5. different

Worksheet 4

1. family
2. good
3. variety
4. Hebrew
5. baramins

Worksheet 5

1. naming
2. species
3. birds
4. taxonomy
5. kingdom

Worksheet 6

1. sorting
2. Creation
3. family
4. tree
5. Mammals

Worksheet 8

1. kind
2. dats
3. species
4. min
5. dogs

Worksheet 9

1. millions
2. 4,300
3. changes
4. baramins
5. family

Worksheet 10

1. hybrid
2. parents
3. Genesis
4. seven
5. 80

Worksheet 12

1. mule
2. striping
3. wholphin
4. Ligers
5. lion

Worksheet 13

1. Latin
2. genus
3. species
4. observed
5. recipe

Worksheet 14

1. dog
2. domestic
3. 500
4. Romans
5. Darwin

Worksheet 15

C	E	L	E	P	H	A	N	T	E
A	M	D	I	N	O	S	A	U	R
N	F	B	Z	U	R	Z	O	S	B
A	P	R	I	N	S	S	D	O	E
R	Y	I	U	T	E	W	Y	B	A
Y	P	N	K	A	U	R	I	E	G
E	S	G	T	I	G	E	R	A	L
R	A	B	B	I	T	L	V	G	E
P	A	P	E	R	V	C	X	L	D
G	O	L	D	F	I	S	H	E	S

Worksheet 16

1. genetics
2. short
3. taming
4. scraps
5. 4,000

Worksheet 17

1. multiply
2. 127
3. Babel
4. orchid
5. variety

Worksheet 18

1. Dogs
2. quick
3. changes
4. double
5. obedience

Worksheet 20

1. China
2. herding
3. wolf-like
4. guard
5. hunting

Worksheet 21

1. grolar
2. white
3. bher
4. courage
5. carnivore

Worksheet 22

1. Noah
2. adapt
3. food
4. creation
5. designed

Worksheet 24

1. Answers will vary.
2. Answers will vary.
3. Answers will vary.

Worksheet 25

1. variation
2. gene
3. blood
4. children
5. reproduce

Worksheet 26

1. 7,000
2. three
3. young
4. vegetarians
5. Meat

Worksheet 28

1. zoos
2. troughs
3. tons
4. movement
5. hibernated

Worksheet 29

1. kind
2. 300
3. 1,400
4. Cross
5. door

Worksheet 30

1. kinds
2. mammals
3. sail
4. pigs
5. canid

Worksheet 32

1. wild
2. Bovine
3. born
4. Flood
5. whale

Worksheet 33

1. beak
2. sauropod
3. headgear
4. four
5. okapis

Worksheet 34

1. 4,200
2. camelids
3. lions
4. extinct
5. crocodile

Worksheet 36

1. spikes
2. plates
3. rhinos
4. tusks
5. Chalicotheres

Worksheet 3

1. Anhinga

2. Double-crested Cormorant

3. Their beaks – the cormorant has a curved tip at the end and the Anhinga doesn't.

4. Answers will vary, but can be something similar to: they both eat fish, live near water, have feathers, beaks, excellent fliers, perch on trees, have bulky bodies, both have special features that allow them to swim and catch fish in the water, both can be water-logged and turn their backs to the sun on land to dry their feathers.

Worksheet 6

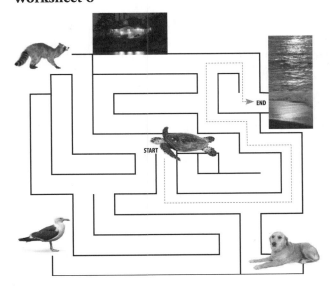

Worksheet 10

1. Answers will vary, but should include that there are two rows of teeth, longer pointed ones on the inside and shorter sawtooth ones on the outside.

2. Answers will vary, but should include that it is a light silver that helps to hide or camouflage the fish in the changing light of the shallow part of the sea. This both protects the fish and makes it easier for it to catch other fish.

3. Answers will vary, but should include that is long (up to 6 feet), large (can weigh up to 100 pounds), and designed to be fast – whether in catching food or escaping from other predators.

4. Answers will vary, but should include that before the Fall, the great barracuda's teeth would have been used to eat a lot of marine plants like giant kelp or sargassum weed. After the Fall, they were used to catch and tear apart fish to be swallowed

Worksheet 11

FOR IN SIX DAYS THE LORD MADE HEAVEN AND EARTH, THE SEA, AND ALL THAT IN THEM IS. EXODUS 20:11

Worksheet 13

M	K	L	A	U	P	S	I	D	E	D	O	W	N	J	E	L	L	Y	D	U	W	Z	K
K	L	G	L	H	G	P	H	A	R	B	O	R	S	E	A	L	B	N	Q	S	J	B	I
M	O	S	L	P	B	K	P	F	L	Y	C	R	T	J	P	J	Z	W	R	W	O	F	L
I	C	E	I	E	E	W	A	W	U	K	D	O	N	O	C	N	G	N	A	N	Z	C	L
D	E	A	G	L	L	P	C	L	B	K	J	H	R	S	E	A	T	U	R	T	L	E	E
C	A	A	A	E	U	G	I	M	X	J	L	U	K	A	P	R	W	G	Y	H	D	P	R
A	N	N	T	W	G	G	F	S	V	W	H	A	L	E	L	F	T	W	Z	C	Q	V	W
W	D	E	O	M	A	O	I	J	N	X	V	O	L	O	G	G	E	R	H	E	A	D	H
B	J	M	R	R	G	S	C	A	M	V	C	D	R	I	K	F	U	L	C	D	H	H	A
F	K	O	D	O	L	P	H	I	N	H	O	R	S	E	S	H	O	E	C	R	A	B	L
K	I	N	I	Y	F	S	U	R	Z	T	B	C	E	B	O	T	T	L	E	N	O	S	E
W	A	E	W	B	E	H	E	S	V	O	F	A	Z	U	H	G	H	D	U	N	N	S
K	P	U	U	Y	D	V	Q	Q	O	N	H	C	S	P	B	X	X	R	O	G	P	Q	W
X	Y	Y	C	Z	O	O	Y	S	Y	X	R	M	T	D	L	G	U	P	O	G	V	U	L
T	F	B	M	O	O	N	J	E	L	L	Y	X	A	O	W	B	S	E	A	L	I	O	N
M	J	W	Q	N	M	K	U	A	L	Z	R	T	R	F	P	B	B	S	M	V	X	S	Y
G	P	A	T	L	A	N	T	I	C	H	L	Y	W	A	L	U	G	M	P	T	W	F	U
F	V	W	S	C	N	X	O	B	E	N	C	H	F	F	R	Z	S	Y	U	V	F	D	W

Worksheet 14

1. 200

2. 15

3. 2 to 3

4. Day 5

5. Answers can include zebrafish, turkeyfish, butterfly cod, and peacock lionfish.

Bonus Question

Answers will vary, but should include one or more of the following: protection/defense against predators, camouflage/hiding/sneaking up on prey, "herding" prey to areas where they can be caught more easily.

Worksheet 17

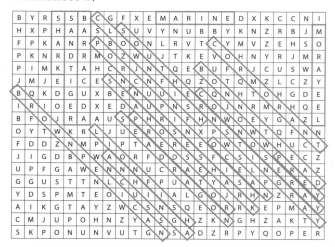

Worksheet 23

Seahorse, Stingray, Lionfish, Bonnethead shark, Leafy sea dragon, Clownfish, Barracuda, Nurse shark, Green moray eel, Sand tiger shark

Worksheet 25

Sea anemone, Bottlenose dolphin, Cownose ray, Penguin

Worksheet 27

Left to right, top to bottom: Seahorse, Leafy sea dragon, Anhinga, Brown pelican, Giant Pacific octopus, Cormorant

Worksheet 31

Answers will vary.

Worksheet 32

Same — swim and feed in ocean; give birth on land; females smaller than males; have whiskers (vibrissae); **Sea Lion** — external ear flaps; streamlined body; solid color fur; four large flippers; **Seal** — no external ears; round body; spotted fur; four short flippers.

Worksheet 1

1. charm
2. 1,260
3. you
4. torpor
5. hover
6. durante
7. constellation
8. short
9. backward
10. bugs

Worksheet 2

1. drumming
2. stations
3. teeth
4. descent
5. grubs
6. murder
7. owls
8. 40,000
9. 50
10. injured

Worksheet 3

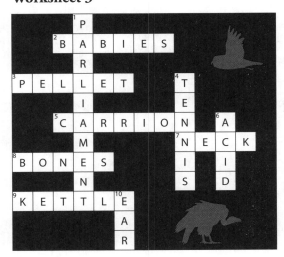

Worksheet 4

1. fall
2. 25
3. feathers
4. backs
5. tip
6. colony
7. meals
8. nests
9. salt
10. cool

Worksheet 5

1. tooth
2. gaggle
3. migrate
4. 70
5. foxes
6. wedge
7. bone
8. eat
9. hollow
10. feathers

Worksheet 6

1. tango
2. gravel
3. color
4. bones
5. dance
6. 10
7. pouches
8. air
9. pod
10. low

Worksheet 7

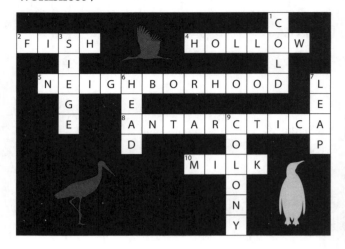

Crossword grid:
- 2 Across: FISH
- 4 Across: HOLLOW
- 5 Across: NEIGHBORHOOD
- 8 Across: ANTARCTICA
- 10 Across: MILK
- 1 Down: COLLIDE
- 3 Down: SINGE
- 6 Down: HEED
- 7 Down: LEAP
- 9 Down: COLONY

Worksheet 8

1. mood
2. Turkish
3. rafter
4. round
5. feather

Worksheet 9

1. poisonous
2. instar
3. 2,485
4. sun
5. swarm
6. group
7. air
8. rest
9. eggs
10. molt

Worksheet 10

1. abdomen
2. pollen
3. hive
4. scents
5. water
6. colony

7. Israel
8. Solitary
9. stab
10. tunnel

Worksheet 11

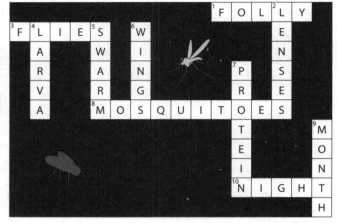

Crossword grid:
- 1 Across: FOLLY
- 3 Across: FLIES
- 8 Across: MOSQUITOES
- 10 Across: NIGHT
- 2 Down: LENSES
- 4 Down: LARVA
- 5 Down: SWARMING
- 6 Down: WING
- 7 Down: PROTEIN
- 9 Down: MONTH

Worksheet 12

1. swarm
2. sclerites
3. blood
4. cut
5. 150
6. colony
7. saliva
8. queen
9. mushrooms
10. west

Worksheet 13

1. army
2. golden
3. day
4. bite
5. less
6. turn
7. parasites
8. 100

9. jaws

10. sleep

Worksheet 14

1. congregation
2. China
3. cow
4. algae
5. protect
6. 10
7. crocodile
8. bank
9. tongue
10. trees

Worksheet 15

1. cold
2. water
3. vulnerable
4. 25
5. mess
6. clamps
7. lounge
8. cells
9. pack
10. mood

Worksheet 16

1. jaws
2. rhumba
3. ear
4. venom
5. toxin

Worksheet 17

1. fangs
2. herd
3. knives

4. wax
5. antlers
6. dromedary
7. China
8. chest
9. sand
10. flock

Worksheet 18

1. mammals
2. 300
3. 3
4. herd
5. dirt
6. 500
7. band
8. grooming
9. ancestor
10. 30

Worksheet 19

1. droppings
2. chins
3. nest
4. June
5. smell
6. dead
7. joey
8. bee
9. claws
10. pouch

Worksheet 20

1. mammal
2. screawa
3. worms
4. territory
5. venom

6. vegetarians

7. nest

8. fur

9. shoes

10. outer

Worksheet 21

1. scurry

2. dray

3. tails

4. 110

5. wood

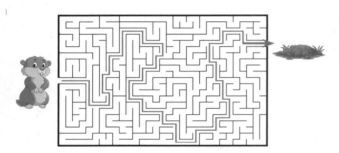

Worksheet 22

1. 6

2. teeth

3. tails

4. dams

5. lodge

6. pig

7. muscles

8. soft

9. backs

10. family

Worksheet 23

1. spray

2. surfeit

3. warnings

4. owls

5. farmers

6. Powhatan

7. rabies

8. nocturnal

9. kits

10. rinse

Worksheet 24

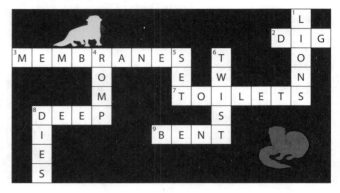

Worksheet 25

1. ferocious

2. boogle

3. speed

4. fur

5. dances

6. tails

7. cat

8. laws

9. mob

10. 20

Worksheet 26

1. rabies

2. leash

3. fur

4. smell

5. prey

6. five

7. legs

8. jaws

9. together

10. pack

Worksheet 27
1. 50
2. pride
3. grasslands
4. social
5. mane
6. cats
7. India
8. solitary
9. ambush
10. burying

Worksheet 28
1. 1,500
2. predators
3. sleuth
4. three
5. hibernation
6. mammals
7. fruit
8. third
9. colony
10. echoes

Worksheet 29
1. 31
2. 100
3. delphis
4. navigating
5. pod

Worksheet 30
1. songs
2. bielo

3. herd
4. blubber
5. warmer

Worksheet 31
1. salt
2. springtime
3. hatched
4. winter
5. hatch

Worksheet 32
1. predatory
2. eat
3. heartbeat
4. shiver
5. Antarctica

Worksheet 33
1. organ
2. temperature
3. yellow
4. shed
5. fry

Worksheet 34
1. backbone
2. stings
3. swarm
4. tissue
5. heart

The Complete Zoo Adventure ⚯ Worksheet Answer Keys

Worksheet 6

```
T U N D R A F C C A D K A N G A R O G H
U R A B V T L H P X T U Z M S Z S M W J
N A B D N A A R N O G A R D O D O M O K
D I D Q V I M I L Z K I B D X I Q L A
G N O I B E I S K A L M I G R A L W F N
L F X W I B N T M S Q O Z V A C O E H G
H O Z E O G N I M A L F X V D V P R U A
O R Q A M A C A C H A P A R R A L T M R
F E M N D N W R N L W B E L R L I Y M O
R S J O U R N A L D O O Z I L L O U I O
O T E N K M E T W A C A M N G I D I N C
S S C S L Q W A T E R P C J H G R Z G H
T Z A F R O G S K F E E V B J A T O B R
R T M G Z O C Y B R R W S C I T Y P I S
E Y E H X C H I M P A N Z E E O U A R T
S U L J C V A I V G T E B X D R M S D S
E S O P R U P O C H T N A H P E L E K Q
D I F P O L A R B E A R N Z K B I D R A
P O A S D C R E A T O R M A L N O F L Z
E R U T N E V D A J Y N O I T A R G I M
```

Worksheet 10

Answers will vary.

Worksheet 14

```
B A T Q W E R C L I M A T E T E L G A E
A H Z X C B K C O C A E P N G I S E D G
S A L O W A L L I R O G L D F Y U I O R
A B T N A L S S A R G M A K Q L M L P A
W I Y V Q X L A N D G B N W E I O K H S
O T D E S I A R A R A T T R T Y U O J S
P A N D A L W O K P A R E O I O N A D L
M T U C W K O A L A S V F L R P T B R A
R E I X G I R A F R D C X G O R A C I N
H P E A G N A K W A H C I B A H I P N D
I L O R E N M D N I K R Z O V F N G O S
N G I R A F F E Z E B O O L F L O A C J
O O A O N O I T A E R C C N R O U R E H
C R D Z N T A E R C L O M A U D S D R P
E R F E R L O W L A N D B M I C G E O A
R I G B T H Y P A X S I V F T L F N S R
O H G R D A N L H G F L I O B I D S A R
S E B A C R E A T I D E G X A R Y H Z O
A Q G N A P T A K R E E M B T M X C V T
Z X C H I P P O P O T A M U S A B N M C
```

Worksheet 17

Answers will vary.

Worksheet 18

Answers will vary.

Worksheet 20

Answers will vary.

Worksheet 22

Answers will vary.

Worksheet 24

Answers will vary.

Worksheet 25

1–6. Day 5

7. Days 5 and 6

8–27. Day 6

Worksheet 29

1. Genus: *Phoenicepterus*, Species: *ruber*

2. Genus: *Pavo*, Species: *cristalus*

3. Genus: *Arahilochus*, Species: *calubris*

4. Genus: *Ara*, Species: *ararauna*

5. Genus: *Aquila*, Species: *chrysaetos*

6. Genus: *Nyctea*, Species: *Scandiaca*

7. Genus: *Pteropus*, Species: *giganteus*

8. Genus: *Canis*, Species: *lupus arctos*

9. Genus: *Ursus*, Species: *maritimus*

10. Genus: *Ailuropuda*, Species: *melanolcuca*

11. Genus: *Phascolarctus*, Species: *cinereus*

12. Genus: *Macropus*, Species: *rufus*

13. Genus: *Panthera*, Species: *leo*

14. Genus: *Suricata*, Species: *suricata*

15. Genus: *Procavia*, Species: *capensis*

16. Genus: *Pan*, Species: *troglodytes*

17. Genus: *Gorilla*, Species: *gorilla*

18. Genus: *Giraffa*, Species: *camelopardalis*

19. Genus: *Equus*, Species: *grevyi*

20. Genus: *Camelus*, Species: *bactrianus*

21. Genus: *Loxodonta*, Species: *africana*

22. Genus: *Diceros*, Species: *bicornis*

23. Genus: *Hippopotamus*, Species: *amphibius*

24. Genus: *Varanus*, Species: *komodoensis*

25. Genus: *Alligator*, Species: *mississippiensis*

26. Genus: *Geochelone*, Species: *elephantopus*

27. Genus: *Pyllobates*, Species: *terribilis*

Worksheet 30

1–6. None

7. Nocturnal

8–9. None

10. Browser

11. Nocturnal/Aboreal

12. Grazer

13. Nocturnal

14. None

15–16. Aboreal

17–18. Browser

19. Grazer

20. Browser

21. Browser/Grazer

22–23. Grazer

24–25. None

26. Browser

27. Aboreal

Worksheet 31

1–6. Bird

7–23. Mammal

24–27. Reptile. Answers to the definitions will vary.

Worksheet 32

Grassland: Kangaroo, Lion, Hyrax, Chimpanzee, Giraffe, Zebra, Elephant, Rhinoceros,

Hippopotamus, Komodo dragon, Tortoise

Hardwood (Deciduous) Forest: Hummingbird, Eagles/Hawks, Fruit bat, Panda, Koala, Komodo dragon, Alligators/Crocodiles

Conifer (Evergreen) Forest (Taiga): Owls, Wolf

Tropical Rain Forest: Flamingo, Peacock, Parrot/Macaw, Fruit bat, Chimpanzee, Gorilla, Elephant, Rhinoceros, Alligators/Crocodiles, Tree frog

Desert: Meerkat, Hyrax, Camel

Tundra: Owls, Wolf, Polar bear

Chaparral: Hummingbirds

Worksheet 35

1. Algae, diatoms, aquatic vertebrates, brine flies, mollusks, shrimp

2. Seeds, fruit, plants, small animals from insects to mice

3. Nectar and insects

4. Tree top seeds, nuts, fruits and berries

5. Squirrels, rabbits, other birds, lizards, and tortoises

6. Lemmings, ducks, other birds, and other animals such as mice and rabbits if available

7. Fruit juice, mangos, papayas, bananas

8. Arctic rabbits, musk ox, caribou, and lemmings

9. Seals, walruses, lemmings, waterfowl, salmon, vegetation grass, sedges, lichens, moss, berries, and seaweed

10. Bamboo shoots plus carrots and protein biscuits

11. Eucalyptus leaves

12. Grasses, some forest plants, leaves

13. Antelopes, zebras, large animals, snakes, insects, nuts/fruits

14. Insects, small animals, birds, roots

15. Vegetation, insects

16. Fruit, insects, honey, meat

17. Fruits, leaves, stems

18. Leaves and twigs of acacia and mimosa trees

19. Grass

20. Plants, fish, flesh, bones, skin

21. Roots, leaves, fruits, grasses, and bark

22. Woody growth and legumes

23. Grasses

24. Deer, goats, wild boar, birds, and other reptiles

25. Fish, turtles, insects, birds, small and large mammals

26. Plants; grass and leaves, cactus, lichen, and carrion

27. Insects

Worksheet 37

1. 5 feet or 55–65 inches, 6–7 pounds, 20 years

2. 84 inches, 6–8.8 pounds, 20–25 years

3. 4 inches, ⅟₁₀ ounce, 5 years

4. 14–33 inches, 1 pound, 40–50 years

5. 30–35 inches, 6–13 pounds, 15–20 years

6. 20–26 inches, 58–61 ounces, 15 years

7. 8 inches, 2.5 pounds, 9 years

8. 40–60 inches, 175 pounds, 8–16 years

9. 4 feet, 1,600 pounds, 25 years

10. 5 feet, 400 pounds, 40–80 years

11. 24 inches, 26 pounds, 10 years

12. 5–7 feet, 65–100 pounds, 12–18 years

13. 5–9 feet, 440 pounds, 10 years

14. 20 inches, 2–5 pounds, 10 years

15. 2 feet, 11 pounds, 7 years

16. 5.5 feet, 100–175 pounds, 40–50 years

17. 6 feet, 600 pounds, 30 years

18. 19 feet, 4,200 pounds, 28 years

19. 8.4 feet, 880 pounds, 10–25 years

20. 7 feet, 1,300 pounds, 50 years

21. 6–12 feet, 14,000 pounds, 70 years

22. 7–14 feet, 3,000 pounds, 50 years

23. 11¼ feet, 5,300 pounds, 45–50 years

24. 10 feet, 350 pounds, 50 years

25. 18 feet, 450–500 pounds, 50 years

26. 5 feet, 600 pounds, 150–200 years

27. 2 inches, 1.8 ounces, 14 years

Quiz 1

Across:

2. Beluga
4. Bottlenose
5. Killer
6. Pacific
11. six
13. Anhinga
14. Loggerhead
16. Pelican
17. Star
18. Octopus
20. Aquarium
21. Nurse

Down:

1. Jellyfish
3. Alligator
7. Atlantic
8. Turtle
9. Stingray
10. Five
11. Seadragon
12. Seahorse
15. Porcupine
19. Shark

Quiz 2

Hooves: 18–23
Pouch: 11–12
Fur/Hair: 7–23
Scales: 24–26
Feathers: 1–6
Claws: 7–17, 24–26
Number of Legs:
2: 1–7, 12, 16, 17
4: 8–11, 13–15, 18–27
Wings: 1–7

Daily Lesson Plan

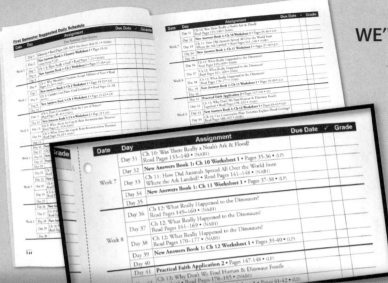

WE'VE DONE THE WORK FOR YOU!

PERFORATED & 3-HOLE PUNCHED

FLEXIBLE 180-DAY SCHEDULE

DAILY LIST OF ACTIVITIES

RECORD KEEPING

"THE TEACHER GUIDE MAKES THINGS SO MUCH EASIER AND TAKES THE GUESS WORK OUT OF IT FOR ME."

★★★★★

HOMESCHOOL

Master Books® Homeschool Curriculum

Faith-Building Books & Resources
Parent-Friendly Lesson Plans
Biblically-Based Worldview
Affordably Priced

Master Books® is the leading publisher of books and resources based upon a Biblical worldview that points to God as our Creator.

MASTERBOOKS.COM
— *Where Faith Grows!* —